A GALAXY OF PRAISE FOR "THE WALTER CRONKITE OF THE SCIENCE BEAT."

D0018183

About the Author

Walter Sullivan, science editor of *The New York Times*, joined the newspaper as a copy boy in 1940 immediately following his graduation from Yale University.

He has filed reports from China, France, and Germany in the fifties and from six expeditions to the Antarctic. He has received numerous awards for his scientific journalism, including the honorary degree of Doctor of Humane Letters from his alma mater and the Distinguished Public Service Award of the National Science Foundation.

His book *We Are Not Alone*, published in 1964, won the International Nonfiction Book Prize. His other books include *Quest for a Continent*, on the Antarctic; *Assault on the Unknown*, on the International Geophysical Year; *Continents in Motion*; and two children's books, *White Land of Adventure* and *Polar Regions*.

BLACK HOLES

The Edge of Space
The End of Time

Walter Sullivan

WARNER BOOKS

A Warner Communications Company

WARNER BOOKS EDITION

This Warner Books Edition is published by arrangement with Anchor
Press/Doubleday, 245 Park Avenue, New York, N.Y. 10017

Permission has been granted by Random House, Inc., for use of the excerpt, appearing
on pages 97–101, from *The Changing Universe*, copyright © by John Pfeiffer.
Permission has been given by the University of Chicago Press for use of the excerpt,
appearing on page 122, from *Quasi-Stellar Sources and Gravitational Collapse*, Ivor
Robinson, et al., copyright © 1965 by the University of Chicago.

The following publications and publishers have given permission for the use of
illustrations:
Astronomy (appearing on page 15), *Nature*, (appearing on page 273). New Science
Publications (for the illustration on page 191, adapted from one that first appeared in *New
Scientist*, London, the weekly review of science and technology), and D. Reidel
and Dordrecht, and the respective authors for the illustrations on pages 190, 192, and
204 from *X- and Gamma-Ray Astronomy*.

Sources for additional illustrations are as follows:
AIP Niels Bohr Library (page 64), The Hale Observatories (page 116), Australia's
Commonwealth Scientific and Industrial Research Organization (page 138), United
States Air Force (page 162), Department of Defense (page 186). The Hale
Observatories (page 242), Kitt Peak National Observatory (page 268), and Cerro
Tololo Inter-American Observatory (page 302). The end paper drawing was provided
by the Goddard Space Flight Center of the National Aeronautics and Space
Administration.

Book design by Helen Roberts

Cover photo by Robert Peck

Maps by Andrew Sabbatini

Diagrams by Bob Michaels

Warner Books, Inc.,
666 Fifth Avenue,
New York, N.Y. 10103

W A Warner Communications Company

Printed in the United States of America

First Printing: November, 1980

Reissued: September, 1982

10 9 8 7 6 5 4 3

Contents

Acknowledgments

Such is the subject matter of this book that the reader must be introduced to phenomena and concepts far beyond direct human experience. The extent to which this effort has been successful has depended on generous help from many quarters. My wife, Mary, reading the manuscript line by line, has insisted that the path be made as easy as possible for the uninitiated. A number of specialists have checked chapters for factual and conceptual accuracy. None, however, have seen the final revisions and none can be held accountable for errors that may remain. George B. Field, director of the Harvard-Smithsonian Center for Astrophysics, read the entire manuscript, and his colleagues, Riccardo Giacconi, William Liller, and William Press, read chapters in which they figure prominently. The same is true of Geoffrey Burbidge, Frank Drake, Herbert Friedman, Thomas Gold, Allan R. Sandage, Kip Thorne, Beatrice M. Tinsley, and John Archibald Wheeler. Other specialists read less extensive sections. Subrahmanyan Chandrasekhar and Gerald Holton drew attention to pertinent early writings of Albert Einstein. Photographs and diagrams were generously provided by William van Altena, Jocelyn Bell Burnell, the Bohr Institute, S. Chandrasekhar, John Cocke, Frank D. Drake, NASA's Goddard Space Flight Center, E. L. Krinov, Roger Linds, George Michanowsky, Guido Munch, Martin Schwarzschild, Richard Tousey, and Virginia Trimble. The following granted permission for use of their published diagrams: Arthur F. Davidsen, et al., William J. Kaufmann (adapted from his book, *Relativity and Cosmology*, 2nd ed., Harper & Row, 1977), G. K. Miley, A. R.

Sandage, and Harvey D. Tananbaum. My daughter Elizabeth and Theodore Shabad helped with Russian translations. The libraries of Columbia and Yale universities and the California Institute of Technology were particularly helpful, as were the Mid-Manhattan Branch of the New York Public Library and the library's Science and Technology Division. Elizabeth Frost Knappman of Doubleday & Co. made many useful suggestions. Brenda Nicolson worked late nights typing the manuscript. While most of those who figure in the book have doctorates and other titles, these have usually been omitted for brevity.

A Note on
Units and Numbers

In view of the current conversion to metric units in the United States (a conversion already complete in other industrialized nations), such units are used in this book except in special circumstances. The reader, however, need only remember that a centimeter is roughly a half inch, that a meter is slightly longer than a yard, that a kilometer is a little more than a half mile, and that at the high temperatures cited in much of the discussion the Fahrenheit figure is roughly double that given (in the Celsius scale). The precise conversion is to multiply the Celsius figure by nine fifths and add thirty-two.

Regarding numbers, "billions" are used in the American sense, meaning one thousand million.

BLACK HOLES

1
June 30, 1908

Few events in recorded history were more awesome—and perplexing—than a colossal midair explosion that occurred over Siberia at 7:14 A.M. local time, June 30, 1908 (as determined from recent analysis of earthquake records). No other occurrence of its kind has ever been observed, and the various proposed explanations are almost equally difficult to comprehend—or believe.

Sober scientists have been led to consider such far-out possibilities as a naturally occurring nuclear explosion, the fall from the sky of an "antirock," or the impact of a small black hole. The latter would be an object so dense that nothing, not even light, could escape its gravity. Within it, theory predicts, should be a "singularity" where time and space are snuffed out.

This book seeks to lead the reader, in as easy steps as possible, along the route of speculation and discovery that has convinced many scientists that black holes exist and have been detected. It seeks to explain what they imply regarding the nature of space, time, and matter—an argument opening vistas into the most far-reaching concepts ever to enter the human mind. It is the author's conviction that anyone who gazes aloft at night and wonders how it all began, and how it will end, would wish to travel the sometimes rocky road of reasoning that seeks to answer such questions.

The debate over the cause of the Siberian explosion—still unresolved—provides an illuminating introduction to the phenomena that, like characters and clues in a detective story, figure in the chapters to come. Many of them are stranger than science fiction. They relate to extremes far beyond the range of our senses of vision, feeling, hearing, and smell. They concern the very small (the building blocks of matter—atoms and their components) and the very large (great assemblages of stars, or galaxies, and the universe as a whole). They deal with the very fast (movements close to the speed of light) and the very strong (forces that would crush a person to oblivion). They evision material so dense a cupful would weigh a billion tons and they relate to the brightest and most distant things in the universe. Yet for all its strangeness this world of extremes is just as real as such near-to-home phenomena as those responsible for television and jet airplanes.

The Event

Although the 1908 explosion apparently shook the entire planet and disturbed the earth's magnetic field, it occurred in a region so remote that the outside world was unaware of it until long afterward. In the nights that followed, skies in northwest Europe were strangely bright, and London residents phoned the police to ask if the northern part of the city were afire. At six British weather stations devices for recording slight changes in barometric pressure (microbarographs) traced what Sir Napier Shaw reported that year as "unaccounted for" oscillations in air pressure. In Potsdam, Germany, atmospheric shock waves traveling around the world in both directions were recorded. Some suspected a tremendous volcanic explosion, like that of Krakatoa twenty-five years earlier.

It was not until scientists reached the region nineteen years after the event that its catastrophic effect began to become known. It was found that within a radius of thirty to forty kilometers almost all trees had been blown down, their uprooted ends pointing toward the site of the blast. Within a radius of five hundred to one thousand meters from the central point some trees were upright but stripped of bark and branches like telephone poles. Virtually all surface material to a distance of twenty kilometers was charred. It was reported that at distances as great as that between New York and Quebec City or between San Francisco and Eugene, Oregon, horses "could not stand up."

6,400 Kilometers Radius

1,500 Kilometers Radius

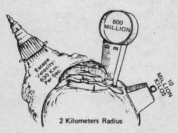

100 Kilometers Radius

2 Kilometers Radius

If the earth were compressed to smaller and smaller size, the weight of all objects would increase rapidly, gravity would become crushingly strong, and escape by rocket would become increasingly difficult. (*Adapted from a diagram by Victor Constanza*, Astronomy)

[15]

Witnesses were few, since the region was sparsely inhabited by Tungus tribesmen—a northern people who subsisted largely by reindeer herding—and by a few Russians exiled by the Czarist government. Among the first visitors were I. M. Suslov, an anthropologist specializing in native peoples of the North, and Leonid A. Kulik from the Museum of Mineralogy of the Soviet Academy of Sciences. Kulik, sent by the academy at the urging of the Russian Society of Lovers of World Knowledge, was, in 1927, the first to reach the site itself. With the help of a native guide, he reconnoitered from Vanavara, a trading station on the Stony Tunguska River, using a raft to probe the north-reaching tributaries. By June he had found the focal point from which the pattern of fallen trees radiated. He noted in his diary, "I still cannot sort out my chaotic impressions. . . . From our observation point no sign of forest can be seen, for everything has been devastated and burned. . . ." It appeared, he said, "that the scorching occurred from the momentary effect of high temperature and not from any ordinary forest fire."

Suslov, traveling on behalf of the Krasnoyarsk Committee for Cooperation with the Peoples of the North, reached nearby communities and heard a succession of hair-raising tales. One told how three Tungus were in their tent forty kilometers from the explosion:

Early in the morning [recorded Suslov] when everyone was aleep in the tent it was blown up into the air, together with the occupants. When they regained consciousness they heard a great deal of noise and saw the forest blazing around them and much of it devastated.

An account that has particularly impressed those who suspect a nuclear explosion (or black hole) was given to Kulik in writing by S. B. Semenov, a farmer at Vanavara sixty-five kilometers from the site:

I was sitting on the porch facing north [he wrote] when in the northwest a fiery blaze appeared for a moment, which sent out such heat it was impossible to stay seated—why, my shirt was almost burned off me. . . . But then the blaze lasted a very short time; I just had time to glance over and see how big it was, and then in a moment it was over. . . . After this it became dark and then an explosion came that threw me from the porch. . . . I was not unconscious very long. I came to, and then this kind of sound came that shook the whole house and nearly moved it from its foundations. It broke the glass and the frames in the houses, and in the middle of the area where the huts stood it tore up a strip of land.

Another resident was facing away from the blast. It felt as though his ears were burning and he clapped his hands over them. Apparently no one was killed, but many reindeer died. A local Tungus, Ilya Potapovich Petrov (Lyuchetkan), said his first wife's uncle had been herding almost fifteen hundred reindeer in the devastated area.

. . . He had many storehouses in this region [according to the account recorded by Kulik] in which he kept clothes, plates, and dishes, reindeer harnesses, and the like. And except for some tens of work animals, all the rest of his reindeer roamed at will in the mountains around the Khushmo River. But down swooped the fire and felled the forest; the deer and the stores were destroyed. The Tungus set out on a search afterward; they found charred carcasses of some of the reindeer; for the rest, they found nothing at all. The fire had burned and melted everything— the clothes, utensils, reindeer harnesses, dishes, and samovars; what they found still left intact were only some buckets.

As the years passed Kulik assembled other reports. One was from Bryukhanov, a farmer who lived near Kezhma on the Angara River several hundred kilometers south of the blast site. At that time he was plowing his land.

When I sat down to have my breakfast beside my plow [he reported], I heard sudden bangs, as if from gunfire. My horse fell onto its knees. From the north

[18]

Trees blown down and charred by the explosion. *(Coutesy E. L. Krinov)*

As many as fifty-six pony-drawn sleighs were used by the 1929 expedition. Here they are shown on the frozen Stony Tunguska River in April of that year.

The Tungus tribesman Ilya Poptapovich Petrov (Lyuchetkan), one of those interviewed by Kulik.

side above the forest a flame shot up. I thought the enemy was firing, since at that time there was talk of war. Then I saw that the fir forest had been bent over by the wind and I thought of a hurricane. I seized hold of my plow with both hands, so that it would not be carried away. The wind was so strong that it carried off some of the soil from the surface of the ground, and then the hurricane drove a wall of water up the Angara. I saw it all quite clearly, because my land was on a hillside.

S. B. Semenov, who was blown off his porch by the blast. *(Photos courtesy E. L. Krinov)*

A river pilot named Kokorin was negotiating the treacherous Mura Rapids on the Angara River, eight hundred kilometers to the southwest. A fiery body considerably larger in appearance than the sun flew across the sky, according to his account (as recorded by Kulik):

Then such a cannonade broke out that all the crewmen who were in the boat rushed to hide in the cabin, forgetting all about the danger that threatened from the rapids. The first bangs were faint, but became progressively louder. The sound effect, he estimated, lasted three to five minutes. The intensity of the sounds was so great that the boatmen were completely demoralized, and it needed much persuasion to make them get back to their places in the boat.

Far to the south a freight train, No. 92 on the Trans-Siberian Railroad, was proceeding from Kansk to Lyalka. The engineer thought an explosion had derailed his train. He stopped, could see no damage, but halted at the next station for a more thorough inspection.

These descriptions were recounted almost twenty years after the fact. Some were secondhand. Others were contradictory as to time, direction, or other details. Since at least some of the Russian investigators did not speak the Tungus language, many of the stories were filtered through interpreters. In search of more immediate accounts the investigators dug into back files of provincial newspapers.

One paper had told of "a noise like the whir of the wings of frightened birds." Another described an "underground roar like the sound of a number of trains passing simultaneously over .rails," followed by fifty or sixty explosions.

Sibir (Siberia), published in Irkutsk, had carried an account by its correspondent in Nizhne-Karelinsk, about a thousand kilometers southeast of the site. He likened the fall to a fiery "pipe" or cyclinder too bright to be viewed directly:

> . . . it seemed to be pulverized, and in its place a huge cloud of black smoke was formed and a loud crash, not like thunder, but as if from the fall of large stones or from gun-fire was heard. All the buildings shook and at the same time a forked tongue of flame broke through the cloud.
>
> All the inhabitants of the village ran out into the street in panic. The old women wept, everyone thought that the end of the world was approaching. In the end it was decided to send a messenger to Kirensk to find out the meaning of this phenomenon.

The Explanations

The earliest visitors to the region assumed that a giant meteorite had fallen. When such an object plunges to earth at very high velocity, the heat generated by its impact causes it to explode, leaving a large crater. The best example is Meteor Crater in Arizona, formed by a meteorite fall between fifteen thousand and forty thousand years ago. It is more than a kilometer wide. The object reponsible for the Siberian catastrophe was long referred to as the Tunguska meteorite, but a succession of expeditions failed to find even a small crater at the focal point of devastation. Instead they uncovered unmistakable evidence that the explosion occured several miles overhead. Tree branches charred by the blast had healed, but when subsequently formed wood was dissected away, scalded wood was found on the upper side of each branch.

Vasily G. Fesenkov, the leading Soviet authority on meteorites, pointed out that such objects rarely hit the earth in the morning. The morning side of the earth faces forward in the planet's orbital flight around the sun, whereas most meteorites overtake the earth, hitting it from behind—from the afternoon-evening side. Such objects are fragments of bodies that, like the planets and asteroids, are circling the sun counterclockwise.

Comets, on the other hand, fly a wide range of orbits around the sun and, Fesenkov noted, were more likely to hit the earth head-on. It is widely believed that the head of a comet conforms to the "dirty snowball" description of Fred L. Whipple of the Smithsonian Astrophysical Observatory in Cambridge, Massachusetts, being formed of frozen gases peppered with dust-sized or sand-sized grains. As such objects swoop into the inner part of the solar system, the most volatile part of this material begins to melt and is blasted off by gas blowing out from the sun (the "solar wind"), forming the comet's tail (which therefore always points away from the sun regardless of the comet's direction of motion).

That the Tunguska object might have been a comet was proposed as early as the 1930s. The brilliant nights observed in Europe following the event, it was suggested, were caused by material shed by the comet head as it plunged into the atmosphere as well as from the comet's tail. Fesenkov estimated that at least a million tons of this material were spread through the atmosphere. The comet head then exploded in midair, not being substantial enough to reach the ground.

Soil samples collected in the area contained microscopic pellets of magnetic iron and silicate globules (although iron pellets were not found in the central region). It was argued that these fragments were what one would expect from the debris of a comet. Such material, however, also rains thinly onto the entire earth from the disintegration of meteors ("shooting stars") in the high atmosphere. Meteors are themselves the debris of comets, meteor showers occurring each year during those periods when the earth passes through the orbit of a disintegrating comet.

Objections, however, were raised to the comet hypothesis. Why, it was asked, was it not seen as it plunged toward the earth? Furthermore, from the observed flight paths of "active" comets (those with tails and a luminous envelope that glows in sunlight), the fall of a comet onto land areas of the earth should occur, on the average, only once in every two hundred million years.

Since no conventional explanation won general acceptance, the field remained open for exotic hypotheses rooted in some revolutionary developments in physics. The latter included the discovery of nuclear energy and the detection of both antimatter and of superdense states of matter including, possibly, black holes.

After the first atomic bombs had exploded over Japan and, later, when a number of high-altitude nuclear-weapons tests had been fired, similarities to the Tunguska explosion became apparent. The latter event and the nuclear detonations shared such features as a brilliant fireball that produced flash burns many miles away, a blast of extraordinary force, and disturbance of the earth's magnetic field. The trees left standing directly below the Tunguska blast, stripped bare like telephone poles, were much like those at "ground zero" beneath the Hiroshima explosion.

Since the physics needed to produce a nuclear explosion was unknown on earth in 1908, other explanations were sought. One proposal, advanced by a Soviet science-fiction writer, Aleksandr Kazantsev, was that visitors from another world were responsible. Kazantsev's thesis, first published in

1946, was that a nuclear-powered spaceship came from Mars to obtain water for that thirsty planet and met with some sort of midair catastrophe. Having visited Hiroshima after World War II, Kazantsev knew nuclear-explosion effects firsthand and he cited the similarities of the Tunguska event. In 1959 a group from Tomsk visited the Siberian site and reported radioactivity in the soil and plants greater than that to be expected from nuclear-weapons fallout, which was then raining on the earth at high intensity.

The Soviet scientific establishment not only dismissed as preposterous the suggestion of alien visitors in a nuclear spaceship, but also was skeptical of evidence for unusual radioactivity reported by "amateur expeditions," although, according to a 1978 account in the British journal *Nature,* the most recent investigations have found a "slight but definite increase" in the radioactivity of surviving trees. A new expedition was planned in 1978 to mark the seventieth anniversary of the event.

Some scientists of impeccable reputation have been reluctant to dismiss out of hand the suggestions of a nuclear explosion—possibly of natural origin. Nonexplosive chain reactions, like those in an atomic power plant, have occurred spontaneously. This happened, for example, in uranium deposits in Gabon, Africa, about 1.8 billion years ago, when the proportion of uranium 235 (the kind used in bombs and reactors) relative to the nonexplosive uranium 238 was sufficiently higher than it is today for chain reactions to occur under special circumstances. That this had taken place was discovered in 1972 by the French, who had found what seemed a deficiency in the inventories of uranium passing through their plant for supplying weapons-grade fuel. They feared theft until they realized that uranium 235 in the mine had been depleted by chain reactions long ago.

Reviewing available evidence on the Tunguska explosion in 1959, Sir William Penney and two colleagues of the British Atomic Energy Authority found much about it to suggest some sort of nuclear detonation. Citing atomic-bomb effects compiled by the United States Atomic Energy Commission, Sir William and his colleagues proposed in the *Philosophical Transactions of the Royal Society* that the extensive blowdown and uprooting of trees must have been

The 1908 explosion occurred over an area (1) north of the Vanavara trading station (2) where buildings were damaged and one man was blown off his porch. At Kezhma (3) soil was blasted off the plowed land and in the Mura Rapids farther down the Angara River (4) raftsmen fled their posts. Near Kansk (5) "horses could not stand up." A freight train traveling from Kansk to Lyalka on the Trans-Siberian Railroad (6) was badly jolted. Near panic was generated in Nizhne-Ilimsk (7) and Nizhne-Karelinsk (8), where some villagers believed the end of the world had come.

caused by an explosion, several miles aloft, yielding about ten megatons. One megaton equals the energy release of one million tons of TNT. Only the larger hydrogen bombs are so devastating. "We guess that the amount of light and heat radiated was less than it would be for a nuclear explosion," they wrote, but they noted that a meteorite entering the atmosphere at the estimated speed of 260,000 kilometers per hour would undergo extreme compression. At a height of 18 kilometers the pressure on its front surface would be 40 tons per square inch (6.5 square centimeters), raising the possibility that extreme compression could have generated a nuclear explosion—for example, in an object rich in heavy hydrogen. In a hydrogen bomb the heavy forms of hydrogen (deuterium

[28]

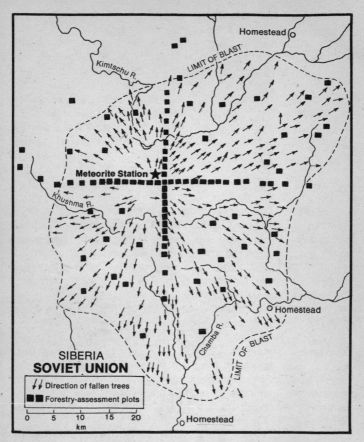

Map of the tree blowdown pattern, from findings of the 1961 expedition.

and tritium) are fused into helium, releasing vast amounts of energy, the pressure and temperature necessary for such fusion being achieved by an atomic explosion based on fission (the splitting, for example, of uranium atoms). A 1976 commentary in *Nature* on this explanation of the Tunguska event said, however, that it was difficult to conceive of a meteorite sufficiently rich in deuterium and tritium. The article also questioned whether entry into the atmosphere could compress and heat the material to the several million degrees needed for fusion and maintain it long enough for an explosion to occur.

In the Soviet Union, Aleksei V. Zolotov of the A. F. Ioffe Physicotechnical Institute of the Soviet Academy of Sciences concluded after working at the site that the blast was like that of a hydrogen bomb (''thermonuclear'') and that vegetation in the area was growing seven or eight times faster than elsewhere, as though stimulated by exposure to radiation. (Kiril Florensky, leader of an eighty-member academy expedition to the site, attributed rapid growth after the destruction to lack of competition from other vegetation, whereas Fesenkov, proponent of the comet thoery, suggested that the ground had been fertilized by organic compounds from the comet head.)

Zolotov argued that the blowdown pattern of fallen trees indicated an extremely concentrated energy source, such as that of a hydrogen bomb, but this, too, was refuted by Fesenkov's group, which simulated the explosion using a ''forest'' of stubby wires three centimeters high capped with cylindrical crowns to represent foliage. From their tests they concluded that the blowdown was caused by the shock wave—a superpowerful sonic boom—of an object descending at an angle of thirty degrees from the ground to terminate in a midair explosion. G. H. S. Jones of the Canadian Defense Research Board has estimated, on the basis of a test in which fifty tons of TNT were detonated over a real forest, that the Tunguska explosion must have been extremely powerful—in excess of two hundred megatons—but was not concentrated enough to be nuclear. (The reason for the Canadian test was not publicized.)

In 1967 Zolotov's argument, based on the blowdown pattern, was published in a prestigious journal of the Soviet Academy of Sciences *(Soviet Physics—Doklady)*. His conclusion, however, that a spaceship was involved has only been aired in the Soviet press and has been ridiculed in *Nature,* which reported that he was recruited for an expedition to the site not for his scientific acumen but because of his knowledge of local conditions as an oil prospector.

The nuclear-blast hypothesis has nevertheless been kept alive by others, notably Ari Ben-Menahem, professor in the

Department of Applied Mathematics at the Weizmann Institute of Science in Israel. He has compared acoustic and seismic data recorded from the 1908 explosion with those generated by Soviet and Chinese nuclear detonations in the atmosphere and has also studied blast effects reported from American weapons tests. The comparisons, he reported in 1975, support the hypothesis that the Siberian explosion had the effects of "an Extraterrestrial Nuclear Missile" with a yield of from ten megatons to fifteen megatons. He referred to it repeatedly as a UFO (for unidentified flying object), although not suggesting it was a spaceship. No single theory, he said, seems to explain all the observations. "We shall perhaps never be able to solve this problem," he added, "unless a similar event reoccurs somewhere on the face of the solid earth." This might resolve the problem, but if it occurred over an inhabited area it could be mistaken for a missile attack, setting off a nuclear holocaust.

While the Russians were experimenting with tree blow-downs, three American physicists assessed a possibility that might have been dismissed as utterly farfetched had not two of them enjoyed worldwide scientific reputations. One, Willard Libby, had won a Nobel Prize in physics—the highest honor bestowed in that field—for his discovery of the carbon 14 "stopwatch" used to determine the ages of prehistoric objects. The other, Clyde Cowan, had been on the two-man team that first observed the ghostlike atomic particles called neutrinos—a major feat of detection. The third was C. R. Atluri, a graduate student working with Libby at the Los Angeles campus of the University of California. Cowan was at the Catholic University of America in Washington, D.C.

In assessing earlier suggestions they said of the comet hypothesis: "It appears unusual that such a comet was not observed on its collision course with the Earth, as it should have been seen unless it approached from a direction with very small angular distance from the Sun."

They considered instead the possibility that the object was an "antirock" formed of antimatter. Here, as with the arguments that have led to the concept of black holes, one is dealing with phenomena that are as real as fire, earth, and water, but that lie beyond the reach of human senses.

Even with our most powerful microscopes we cannot see atoms in the way we see bigger things. Trying to do so, using ordinary light waves, would be like trying to feel the shape of an amoeba or other microscopic creature with your fingers. You cannot do it because your fingers are much larger than the amoeba. Only with very short wavelengths—those of X rays—is it possible to detect individual atoms. We can, however, learn about them through a variety of other means. We can study their properties and predict their behavior in ways that are of enormous practical importance. Our understanding of what goes on at the atomic level has brought us not only nuclear energy but also transistor radios, television, pocket calculators, and the components of an almost unlimited list of other items that fashion our lives.

The atom, we now know, can be satisfactorily described as formed of three kinds of particle: protons and neutrons (clustered in its tiny, very dense nucleus) and electrons (which "orbit" that nucleus). It is electrons that flow through our wires, illuminate our lights, and do thousands of other jobs. While it is misleading to describe them as orbiting the atomic nucleus the way planets orbit the sun, they do occupy clearly defined orbital "slots." Furthermore, the atom can be viewed—like the solar system (the sun and its family of planets)—as being chiefly empty space.

The proton is roughly 1,836.1 times heavier than the electron (an odd and unexplained relationship) and carries a single positive electric charge. The electron, despite its far lighter weight, has an electric charge of equal strength, but it is negative. When an atom has its full complement of electrons, there is one matching each proton; the negative charges thus cancel out the positive charges of the nucleus and the atom is electrically neutral. The neutrons, as their name implies, are themselves electrically neutral. They add weight to the nucleus (and thus to the atom) but they do not affect its chemical behavior, which is determined entirely by electrical forces. Protons and neutrons (which are slighty heavier) have been shown, under bombardment by other particles, such as electrons, to have some sort of internal structure. They are "made of something." Some call these constituents "quarks." The protons and neutrons can be smashed, producing a wide range of short-lived particles but never (so far, at least) have quarks appeared. No one has been able to show that electrons have any "size" at all, and they are treated by physicists as being infinitesimally small.

One of the oddest findings of such research has been that for each of these particles—the electron, proton, neutron, and all other fruits of atom-smashing—there is a sister particle that is opposite in such properties as electric charge or spin. These are the particles forming antimatter.

As so often happens in science, the idea did not burst full-blown from the blue. It evolved step by step. Its father was P. A. M. Dirac, a British theorist whom many rank with Einstein. In 1928, he was seeking to formulate the behavior of matter and energy in ways that were symmetrical, so that nature did not appear to favor positive charge vs. negative charge, or right-handed, clockwise behavior vs. left-handed, counterclockwise behavior. It seemed, to make things orderly, that energy must occur in negative as well as positive form. Furthermore, the implication was that for each of the particles of matter—electrons, protons, neutrons, and so forth—there must be a sister particle that was identical in mass (weight) but opposite in some property such as electric charge.

Four years later, in 1932, it was discovered that, in fact, there is such a mirror image of the electron. Since its electric charge is positive (as opposed to the negative charge of the electron), it was named the positron. Gradually the antimatter counterparts of other well-known particles were discovered—the antiproton, antineutron, and so forth. When such a particle encounters one of matter, the two mutually annihilate one another, becoming converted to energy in the form of gamma rays (extremely high-energy counterparts of light rays).

Because our world is dominated by matter, any particle of antimatter that makes its appearance meets an atom of matter almost instantaneously and is annihilated. This occurs continuously because particles of antimatter are formed when high-energy particles from space (cosmic rays) that steadily rain on the earth collide with atoms in the air or in more solid material to produce sprays of other particles, including some of antimatter. In this fleeting way antimatter is a part of our environment. If our eyes were capable of doing so, we would see tiny gamma-ray flashes occurring constantly all around us as miniature matter-antimatter encounters occur.

Whereas an atomic- or hydrogen-bomb explosion converts only a very small fraction of the fuel (such as uranium) into energy, the conversion is total in an encounter between matter and antimatter. Hence Libby and his two colleagues reasoned that a very small "antirock," when it encountered the atmosphere, would produce a terribly big bang.

Because theorists are devoted to the idea of symmetry, there are those who believe not only that the universe must hold equal numbers of positively and negatively charged particles (so that as a whole it is electrically neutral), but also that there must be equal amounts of matter and antimatter. Perhaps, they say, far out in space, there are galaxies of antimatter stars and planets. They would look the same as our galaxy—the Milky Way—because an antimatter candle or star would produce light that is indistinguishable from that generated by matter.

If such galaxies exist, Libby and his colleagues reasoned, perhaps once in a while a chunk of rock escapes them to fly across the great void to our own galaxy—an antimatter meteorite. If it entered the atmosphere, a horrendous fireball would result, creating the observed flash, the burns, the heat, and the shock waves.

It was known that high-altitude nuclear-weapons tests produce a great deal of carbon 14—the radioactive form of carbon on which Libby had done his prize-winning research. A matter-antimatter explosion should have produced even more carbon 14, and this could have become incorporated into plant life growing during the period immediately following the 1908 blast. In particular it should be in the wood of tree rings formed in 1909.

This type (isotope) of carbon is called carbon 14 because its nucleus is formed of fourteen particles—six protons and eight neutrons. The more common forms of carbon—carbon 12 and carbon 13—also contain six protons, but they have fewer neutrons. That makes them lighter in weight, but it is those six protons that determine the electric charge on the nucleus and therefore the chemical properties of the atom. All carbon atoms—and no others—contain six protons.

[34]

Unlike the other forms, carbon 14 is unstable—that is, it is radioactive and decays at a fixed average rate into nitrogen 14. Half of a given amount decays in 5,730 years, that being its "half life." There would not be any carbon 14 around (apart from that produced by atomic bombs and the like) but for the fact that new carbon 14 is constantly being made high in the atmosphere by the impacts of cosmic rays. The latter are called "rays" because at first they were thought to be forms of light even more penetrating than X rays. They are, in fact, predominantly atomic nuclei, and they are continually splitting atoms of the upper atmosphere, knocking free their neutrons. Some of the neutrons penetrate nuclei of the atoms that make up most of the atmosphere—those of nitrogen 14. The latter contain seven neutrons and seven protons, but the invading neutron knocks out one of the protons, leaving six protons and eight neutrons—that is, carbon 14. The newly created carbon atom immediately "marries" two oxygen atoms to form carbon-dioxide gas.

Unlike most carbon, however, this form is unstable. Its nucleus is "uncomfortable" with eight neutrons. Normally a neutron, which is essentially a proton wedded to an electron, is quite happy inside a nucleus. Outside, free of the strong forces at work in the nucleus, neutrons are unstable. They decay into protons by firing off an electron (plus an almost undetectable antineutrino), a common form of radioactivity. Inside the nucleus they do not decay except when they are "uncomfortable"—when they do not really belong there, as in the carbon 14 nucleus. Thus eventually one of the neutrons in carbon 14 decays, ejecting an electron and turning the atom back into one of nitrogen 14.

Most of the carbon in living matter has come directly or indirectly from the air and contains a certain percentage of carbon 14. Plants "inhale" carbon dioxide and "exhale" oxygen, helping to replenish the oxygen we breathe. The plants thus form carbohydrates that are eaten by us as vegetables or by the animals we eat. But once we or any other living things die, this replenishment of the carbon ends. There is no longer any input from the atmosphere, and the percentage of carbon 14 in skeletons, dead wood, and so forth relative to the stable forms of carbon decreases at a steady rate. This is the carbon 14 stopwatch. Even at the start, the percentage of carbon 14 is very small. In the atmosphere only about one carbon atom in one thousand billion is carbon 14. But since they are radioactive, small amounts can easily be measured, and so, as Libby demonstrated, carbon 14 could be used to determine the ages of once-living material up to a current limit of about fifty thousand years.

Future generations will have difficulty applying this method to anything that lived after the first atomic bombs went off because the explosions added to the atmospheric carbon 14, and the production rate was no longer uniform. Libby and his coauthors calculated that by 1961 weapons with a cumulative yield of seventy megatons had been fired in the atmosphere and another hundred megatons on the surface. As a result, plants were taking up 25 per cent more carbon 14 than normal. Would they find a similar effect in wood that grew in 1909?

To learn the answer they obtained wood from a succession of annual rings in the "Hitchcock Tree," a three-hundred-year-old Douglas fir that had fallen in a winter storm in the Santa Catalina Mountains of Arizona. The same was done for an oak that had grown in the Simi Valley near Los Angeles.

In both trees they found that 1909 was the only year in which the carbon 14 levels were substantially above normal—higher than what they termed a "reference level" determined from a variety of observations. The excess, however, was less than one seventh what they expected, had the explosion been due entirely to antimatter.

The antirock idea seemed even more doubtful later that year (1965) when it was reported that in a fir tree near the Oregon-California border neither the 1908 growth (the year of the blast) nor a 1912 specimen showed unusually high levels of carbon 14. Following this, however, Robert V. Gentry at the Oak Ridge National Laboratory in Tennessee proposed that Libby and his colleagues had overlooked an effect that would reduce the amount of extra carbon 14 to be expected from a matter-antimatter explosion. Therefore, he said, the findings "are consistent with the hypothesis that the Tunguska meteor was entirely antimatter in content." He also found such a blast a more likely explanation than the hydrogen-bomb type because from American bomb tests it appeared

[36]

that a thermonuclear fireball comparable to the Tunguska explosion would have persisted thirty-three seconds. By contrast, he said, "it certainly seems possible" that an antimatter explosion of this yield would last only a few seconds, conforming to the accounts of eyewitnesses such as Semenov.

Nevertheless, the antirock idea never won wide support, and the way was clear for an even more far-out proposal—that the explosive effect was caused by a "mini black hole." The authors of this suggestion were A. A. Jackson and Michael P. Ryan, Jr., then both at the Center for Relativity Theory of the University of Texas in Austin.

In a black hole, gravity—normally the weakest of the basic forces in nature—becomes a tyrant. As Sir Isaac Newton figured out three centuries ago, the gravitational force exerted by a body, such as the earth, moon, or sun, is proportional to the body's mass. Since the earth's mass is eighty-one times that of the moon, the strength exerted by the earth's gravity in space at a given distance from the earth's center is eighty-one times stronger than the gravity of the moon at a similar distance.

Gravity also weakens at increasing distance from its source, according to the rule that also applies to the brightness of a light. If you double the distance, the brightness or the force of gravity becomes one quarter what it was. (It is called the "inverse square law" because the strength falls off according to the square of the distance.) As you approach the earth, gravity increases at the same rate.

But let us now think in terms that would never have occurred to Newton. Suppose while you were off on a space trip someone squeezed the earth into a volume with half its present radius, making it that much denser. When you reached a point that would have been on the earth's surface before it was squeezed, the force of gravity would be the same as the one we all live with. But when you got to its newly compressed surface, half the distance to the earth's center, its gravity would be four times greater. If the earth had been shrunk to a quarter its radius, its surface gravity would be sixteen times greater (see page 15).

Now for the really big "suppose." Imagine that the entire earth has been squeezed down to the size of a Ping-Pong ball. Then you have all the mass of the earth—billions upon billions of tons—compressed into that tiny volume exerting a gravitational force so strong that—for reasons we shall see later—light could not escape it. The earth would be invisible—a "black hole."

The "conventional" concept of a black hole is of a star that has collapsed to a state of superdensity and become invisible. In 1971, however, Stephen Hawking of Cambridge University in England proposed that there might be "mini" black holes wandering throughout the universe as a residue of the "Big Bang" explosion in which the universe was born. The concentrated, incredibly violent turbulence following that primordial explosion could have compressed concentrations of matter sufficiently to produce black holes of submicroscopic size.

Hawking, although stricken by a progressive nervous disorder (amyotrophic lateral sclerosis), has become one of the most brilliant and innovative theorists on black holes. At scientific meetings the motorized wheelchair that supports his limp frame (he operates it by push-button control) is habitually surrounded by other theorists eager to exchange ideas with this man whose bright, eager face resembles that of a fifteen-year-old. Because his voice is weak they must stoop to hear and must write out his equations for him. His ability to perform long calculations without such assistance has been likened by one colleague to Mozart's composing an entire symphony in his head.

The proposal of Jackson and Ryan was that the earth had encountered a mini black hole with a mass comparable to that of a large asteroid but with dimensions not much larger than those of a single atom. If, as Hawking proposed, mini black holes are wandering here and there through the universe, one of them might have fallen to the earth. Starting with zero velocity relative to the earth, it would have arrived with sufficient velocity to traverse the lowest thirty kilometers of the atmosphere in about one second, creating a shock wave in the air of sufficient intensity to produce temperatures in the range of ten thousand to a hundred thousand degrees. The initially radiated light would be chiefly at invisible ultraviolet wavelengths but would be absorbed and reradiated by air along the path, creating the blindingly blue "pipe" or fiery pillar described by many witnesses. The heat and radiation would

Stephen Hawking in his push-button wheelchair. *(Godrey Argent, Camera Press London)*

have produced the flash burns and charring of wood. The blast would have blown down the trees.

"Since the black hole would leave no crater or material residue," said the authors, "it explains the mystery of the Tungus event. It would enter the Earth, and the rigidity of rock would allow no underground shock wave. Because of its high velocity and because it loses only a small fraction of its energy in passing through the earth, the black hole should very nearly follow a straight line through the earth. . . ."

Using published (but frequently revised) estimates of the direction of its arrival, they proposed that it emerged from the earth in the North Atlantic roughly midway between Spain and Newfoundland. As it did so and flew outward through the atmosphere, it should have created disturbances in the sea and generated a new air shock wave. The authors urged that weather-station records be checked "for an event similar to that caused by the entry shock displaced by the proper amount of time. Oceanographic and shipping records should be studied to see if any surface or underwater disturbances were observed."

It would, indeed, be startling if a mini black hole erupted from the sea close to a passing ship, but there seems to be no record of such an occurrence or of additional atmospheric shock waves. Could it be that a mini black hole would disintegrate in the atmosphere or within the earth instead of passing through? Could it somehow account for the slightly abnormal level of carbon 14 reported by Libby and his colleagues from wood formed the following year? Sir William Penney and his colleagues had suggested that the excess carbon 14 could have originated in a nuclear explosion. In 1977, however, John C. Brown and David W. Hughes of the Universities of Glasgow and Sheffield in Britain argued that the explosion of a comet could generate enough heat for nuclear reactions to produce carbon 14.

What happened on June 30, 1908, remains uncertain, and the debate continues. The Soviet press has kept alive the spaceship thesis of Aleksei Zolotov and in October 1978 reported an interview with another promoter of that theory (Felix Zigel). Almost all scientists, however, prefer the less exotic explanations. In 1978 L. Kresák of the Astronomical Institute of the Slovak Academy of Sciences in Bratislava advanced the idea that, from the little that is known of the direction from which the object approached, it could have been a fragment from Comet Encke, which, with the shortest period of all comets, orbits the sun every 3.3 years. He suggested a chunk about a hundred meters in diameter that was "extinct" (no longer carrying the envelope of volatile, luminous material that is evaporated and blown off to produce the tail of an "active" comet). Such "cometary boulders," he said, probably represent "an overwhelming majority" of interplanetary objects in the size range from one to one hundred meters.

Every year at the time of the Tunguska event, Kresák pointed out, the earth passes through a stream of debris along the orbit of Encke, producing one of the Taurid meteor showers. He conceded, however, the existence of other possibilities, such as a type of meteorite particularly subject to disintegration in the atmosphere (a form of carbonaceous chondrite), as proposed by Fred Whipple.

Whatever the cause of the Tunguska event, debate concerning it has brought to public attention some of the most astonishing developments in modern physics. It has transformed "black holes" into a popular (if mystifying) topic of conversation. For the scientists it has coincided with widespread excitement regarding the evidence that such objects—long considered a theoretical oddity—may in fact occur, providing answers to a whole range of problems. Some astronomers suspect they constitute a large part—perhaps the major part—of our universe. Black holes have been seized upon to explain phenomena ranging from quarks to the extraordinarily violent processes within the hearts of galaxies. This spate of speculation has culminated a succession of discoveries showing that matter sometimes occurs in states far more concentrated than seems plausible from everyday experience.

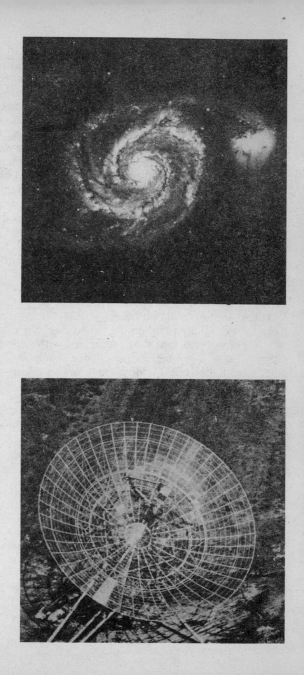

2

Ten Tons
per Cubic Inch

To a remarkable extent, studying the stars has told us things about nature that we could—or might—never have learned had we not looked beyond the earth. Newton's revolutionary discoveries regarding gravity were largely based on observations that described the movements of the moon and planets. Lockyer did not discover helium in the laboratory but in terms of its characteristic wavelengths in sunlight. He called it "helium" from the Greek for "sun." And it was celestial observations that led to the finding that matter can be compressed to almost unbelievable densities.

An early clue came in 1844 when Friedrich W. Bessel at the observatory of Königsberg in Prussia (it is now Kaliningrad in the Soviet Union) observed that Sirius, the brightest of all stars, does not move among the other stars in a straight line but in a slightly irregular manner.

We think of the stars as being fixed relative to one another. The Big Dipper or the constellation Orion seem the same from year to year. But the stars are actually in motion with respect to one another. Over the years and centuries the observed positions of the brightest (and hence nearest) change slightly.

The fact that the motion of Sirius was along a wavy rather than a straight line suggested to Bessel that the star had an invisible companion. It was already well known to astronomers that the earth in its annual flight around the sun does not move in a smooth ellipse. (All orbiting bodies, as Kepler discovered almost four centuries ago, move along elliptical rather than circular paths.) Rather there is a slight waviness to its movement because of the influence of the moon.

While it is the gravity of the earth that holds the moon in orbit, the gravity of the moon, as it orbits, is sufficient to pull the earth to one side or another of the earth's own orbital path around the sun. When the moon is on the side of the earth opposite the sun (full moon), it pulls the earth away from the sun. At new moon, two weeks later, the moon pulls the earth toward the sun. The effect, through each year, is that the earth weaves in and out of the path that it would fly were the moon not there. If a distant astronomer could see the earth but not the moon, he could still infer the moon's presence from this effect.

So was it with Bessel. Astronomers, through their study and classification of stars in terms of those that, for example, are larger or smaller, brighter or dimmer, redder or bluer than the sun, had been able to calculate the mass of Sirius rather accurately. The term "mass" refers to the amount of material in an object, be it anything from a tiny electron to a star like the sun. It is more universally applicable than "weight," which is defined by the response of an object to gravity. Gravity varies from place to place (it is, for example, a tiny bit stronger in Minneapolis than in Chicago), whereas the mass of an object, measured by its inertia (its resistance to acceleration), is everywhere the same, be it on earth or far out in space. The mass of a star is usually defined relative to the mass of the sun. That of Sirius is 2.28 solar masses.

From the wavy motion of Sirius it was possible to determine the time it took its unseen companion to complete one orbit around the star and, from the extent to which Sirius—a gigantic star—was being pulled off its path, it was possible to estimate the mass of the companion. It is roughly equal to that of the sun. Yet even though it is one of the nearest stars to earth, it seemed invisible, which was hard to understand.

Nineteen years after Bessel made his observation Alvan Clark of a famous American telescope-making family found the companion while trying out an eighteen-inch lens that was eventually to be installed at the Dearborn Observatory near Chicago. The companion was extremely dim. When viewed from the same distance as the sun is from the earth it would be only one four-hundredth as bright. Among typical stars the small, dim ones burn at lower temperatures and are redder. The big, hot ones burn white. This one was as white as Sirius—considerably whiter than the sun. Why, then, was it so dim?

The only explanation that astronomers could think of was that its surface area was very small. Yet to exert its observed influence on Sirius it must be extremely massive—and therefore very dense. Material equal in mass to that of the sun must be squeezed into an area no larger than a planet or, in terms of more down-to-earth equivalents, the material of a large stone must be compressed to a grain of sand.

As Sir Arthur Eddington put it in 1927: "The message of the Companion of Sirius when it was decoded ran: 'I am composed of material 3,000 times denser than anything you have come across; a ton of my material would be a little nugget that you could put into a matchbox.' What reply can one make to such a message? The reply which most of us made in 1914 was—'Shut up. Don't talk nonsense.' "

While it seemed nonsense in 1914, two centuries earlier that pillar of modern science, Sir Isaac Newton, considered radical condensations of matter in the nature of things. Newton, of course, was unfamiliar with modern concepts of atomic structure and the forces governing it, but he knew that "solid" matter is transparent to magnetic and gravitational forces and, often, to light. Matter, he wrote in the 1704 edition of his *Optiks,* is "much more rare and porous than is commonly believed." He noted that water, with only one nineteenth the density of gold, is still relatively incompressible. The implication was that resistance to compression derived from structural forces rather than true solidity. In the 1717 edition of that classic work he argued that solid bodies are formed of particles with large empty spaces in between, that the particles themselves are formed of tinier particles, similarly distributed through space, and so on "perpetually" to smaller and smaller levels of matter. Following up on these ideas, his contemporary, John Keill, proposed that the material forming so transparent an object as a glass "has not a greater proportion to its magnitude, than a grain of sand to the whole bulk of the globe of the earth." In another popularization of Newton's theories Henry Pemberton wrote that all bodies of the universe "may be compounded of no greater a portion of solid matter, than might be reduced into a globe of one inch only in diameter, or even less.''

Later in that century Joseph Priestley, the discoverer of oxygen, carried Newton's ideas even further. Not only might all solid matter of the solar system ''be contained within a nutshell,'' he said, but also the concept of solidity might be meaningless—''there might be no such thing in nature.'' One could argue that, in this, he anticipated the concept of a ''singularity''—a contraction of matter (as in the core of a black hole) to infinite density and infinitesimal volume.

With the advent of modern concepts of atomic structure Newton's ideas in this regard had largely been forgotten and so discovery of the companion of Sirius—the first ''white dwarf''—came as a deep shock. With only one known, the evidence was not convincing, but soon more were found. They are, in fact, common, but being dim, only the nearest can be seen.

Meanwhile, by the 1920s, enough had been learned about atoms and enough had been guessed about what makes stars shine to suggest how white dwarfs might come about. Scientists had long since given up the idea that the sun ''burns'' fuel in the ordinary sense. In the nineteenth century it had been proposed by Helmholtz in Germany and Lord Kelvin in England that slow contraction of the sun and other stars, because of their enormous weight, could generate enough heat and radiation to produce the observed energy release. The trouble was that this could keep a star like the sun shining for only a few million years, yet there was ample evidence that the earth was billions, rather than millions, of years old. It was highly unlikely that the earth was older than the sun.

In the early years of this century, however, Albert Einstein had formulated, as a follow-up to his first theory of relativity (the ''special'' theory), a relationship between energy and matter that could be interpreted as opening up stupendous possibilities. It suggested that a very small amount of material was equivalent (in ways then not clearly understood) to a very large amount of energy, expressed by the famous formula $E = Mc^2$. It stated that energy (E) equals mass (M) multiplied by the speed of light (c) multiplied by itself (c^2). Since the speed of light is an extremely large factor, the energy produced, if one could achieve the conversion from mass, would be enormous.

For example, if under the extremes of temperature and pressure in the core of the sun four hydrogen nuclei were fused to form a single helium nucleus the latter would be 0.8 per cent lighter in weight than the original four nuclei, the difference being in the amount of nuclear "glue" (binding energy) in the helium nucleus compared to that in the four hydrogen nuclei. The 0.8 per cent residue would be converted to much energy. (This, as noted earlier, is far less efficient than the matter-antimatter reaction in which all the mass of the interacting particles is converted to energy.)

It was not until a number of years after Einstein's formulation of the energy-mass relationship that its significance as a potential source of energy on earth was recognized. And it is typical of such "new" ideas that the conversion of mass to energy and vice versa had been speculated upon by Sir Isaac Newton just two centuries earlier. In his *Optiks,* discussing light (as a form of energy), he wrote: "Are not gross bodies and light convertible into one another; and may not bodies receive much of their activity from the particles of light which enter into their composition?"

Einstein's formulation differed, of course, because the speed of light was a factor. While several years passed before its validity was demonstrated in the laboratory, leading ultimately to the atomic and hydrogen bombs and atomic energy (more properly "nuclear" energy), astronomers felt they were on the track of what makes stars shine. Since most stars are chiefly formed of hydrogen, by fusing it into helium they could "burn" for billions of years (it is estimated that the sun converts 100 million tons of hydrogen into helium every second, but still has enough hydrogen left to last several billion years).

The other new knowledge that suggested how white dwarfs might come about concerned the nature of atoms. The Dane, Niels Bohr, in 1913 proposed a model of the atom that, while it has since been considerably modified, helped lay the basis for modern theory. It must be remembered that when we talk about atoms and the particles of which they are formed we are discussing things no one can see. To describe them in terms of familiar things is bound to be imprecise. They behave according to rules completely unfamiliar to us in our everyday lives. Another civilization could describe atoms and their behavior in quite different ways and be just as accurate. And because new discoveries are continuously being made in physics, our descriptions keep changing too.

Nevertheless, the "Bohr atom," as it came to be known, was a good starting point for efforts to understand how matter could be greatly compressed. As noted earlier, the atom, in this sense, can be likened to the solar system, with electrons as planets and a tightly packed nucleus of much heavier protons and neutrons as the counterpart of the sun. While atoms are largely empty space, they can form substances that feel very solid because of the electrical forces that bind them together.

The simplest atom of all is that of hydrogen, with but a single proton (plus, rarely, one or two neutrons) as its nucleus and one electron in orbit. The heavier atoms have nuclei formed of many protons and neutrons and have numerous orbiting electrons. The latter are distributed through successive orbital "shells," with room for two electrons in the first, innermost shell, eight in the second, eighteen in the next, and so forth. The electrons within any shell (apart from the innermost) can move at one of several energy levels. Finally, they do not "orbit" in a predictable manner—an aspect of "quantum" behavior that is a basic feature of everything that happens on the atomic level.

There is no need here to plunge deep into the mysteries of quantum behavior, but a few of its aspects are critical. The term "quantum" comes from the discovery that phenomena involving atoms do not happen in a gradual, continuous manner, like lights being turned up or down in a theater. They occur in jumps. When electrons change their energy state—as in changing orbits—they give off or absorb energy in distinct little packets or "quanta," and this quantum behavior permeates all phenomena relating to atoms and molecules.

One aspect of such behavior is a form of "snobbishness" on the part of electrons that eventually helped explain the nature of white dwarfs. This is the "exclusion principle" set forth in 1925 by Wolfgang Pauli, an Austrian physicist, which specifies that no two electrons in an atom can share all of the same four basic properties. These relate to:

1. Which orbit the electron occupies.
2. Which energy level it occupies within that orbit.
3. The geometry of its orbit.
4. The direction (clockwise or anticlockwise) of its "spin."

The atom can thus be likened to a rather strange tennis club that allows two players on the first court (the innermost orbital shell), eight on the second court (the second shell), and so forth. While the players on these courts are in constant motion, each one on the average plays a specific position and does so alone. No other members can play there at the same time.

That electrons retain this snobbishness even when they are not orbiting a nucleus was recognized by Enrico Fermi, the Italian physicist who, in a University of Chicago squash court, directed the project that, on December 2, 1942, achieved the first sustained chain reaction, ushering in the atomic age. Because that was the top secret project designed to produce the first atomic bomb, word of Fermi's success on that historic day was passed to the president of Harvard University, James Bryant Conant (himself a participant in the effort), via a cryptic telephone call that became famous. "Jim," said the caller (Arthur H. Compton), "you'll be interested to know that the Italian navigator has just landed in the New World."

Fermi found, in his electron studies, that when electrons are moving freely, as in a very hot gas or plasma, those with identical properties do not wish to associate with one another. They resist compression with what has come to be called "electron pressure."

The key role of such pressure, on the road to the black-hole concept, was not immediately recognized. An initial step was the realization that atoms, in terms of their electron structure, are largely empty space and that it should therefore be possible, under great pressure, to squeeze matter into an extremely dense state—if it is cold enough. In a very hot gas, like that forming the sun and other active stars, the electrons are heated to such high temperature—and are therefore moving so fast—that they break free from atomic nuclei, forming a "plasma." Because of their motion they exert an outward pressure, such as that which causes any heated gas to expand, if free to do so. This is what happens in the cylinder of a car or steam engine. It is such thermal energy plus radiation flowing out from the core of a living star that keeps it from collapsing under its own weight.

In 1926 Ralph H. Fowler in Britain saw that, when a star has finally burned up its nuclear fuel, the outward pressure must cease and the star collapse to extreme density much, he said, like "a gigantic molecule in its lowest quantum state." It is as though the steel framework of a great skyscraper suddenly turned to rubber and the building collapsed into a pile of rubble.

But what was it that then halted the collapse? What determined how big a pile of rubble would remain—the size typical of a white dwarf? And was this the fate of all stars that have burned up their fuel?

It was a nineteen-year-old Indian named Subrahmanyan Chandrasekhar—now known to astronomers throughout the world as "Chandra"—who worked out the answers to these questions. As often happens in science, Chandra had been born to a tradition of discovery. His uncle, Sir Chandrasekhara Venkata Raman, had just identified the "Raman effect" in light scattering that immortalized his name and won for him a Nobel Prize in physics. Young Chandra worked out his solution to stellar collapse during the idle hours of a long voyage from India through the Red Sea, the Suez Canal, and the Mediterranean to Venice, en route to do graduate work at Cambridge under Fowler, the man who had proposed the collapse idea.

Only a year or so earlier Fermi and Dirac had presented a statistical way to describe how free electrons would behave according to the snobbish rules of Pauli's new exclusion princple. Chandra presented his first analysis in 1930 (he revised it somewhat two years later, taking into account an important effect, namely that when so compressed, electrons would be moving so nearly at the speed of light that relativity would play an important role). The key element was that, from the

Subrahmanyan Chandrasekhar

[51]

Like a classroom assemblage of very exceptional students, a number of those who laid the foundations of modern physics attend a 1930 meeting at Niels Bohr's institute in Copenhagen. In the front row, from left to right, are Oskar Klein, Bohr, Werner Heisenberg, Wolfgang Pauli, George Gamow, Lev Landau, and Hendrik Kramers. Edward Teller is looking over Kramers' shoulder. The toys on the counter were reportedly placed there by Bohr as a prank: a trumpet to enable Pauli to "blow his own horn," a cannon so that Gamow could "shoot down" foolish hypotheses, and a toy soldier in front of Kramers. Those in the rear rows are Ivar Waller (between Klein and Bohr), Piet Hein, Oscar K. Rice, Rudolph Peierls, Aurel Wintner, Walter Heitler, Christian Moller, Felix Bloch, Mogens Pihl, Tanya Ehrenfest (looking out from behind Landau), Hansen (identification uncertain) behind her, and William Colby. The photograph was made at a somewhat lighthearted session on the structure of the electron. Heisenberg reportedly proposed that its color must be yellow. "If you see green," he explained, "you go ahead; if you see red, you stop; but if you see yellow you don't know what to do." A half-century later the electron still seems to behave as though it were a point with no structure at all. *(Niels Bohr Institute)*

Fermi-Dirac statistics, it was clear the "electron pressure" could resist only a limited amount of compression. If, as Chandra finally worked it out, a collapsing star was more than 1.4 times as massive as the sun, electron pressure could not stop its collapse.

The maximum mass of a white dwarf was therefore 1.4 solar masses. Yet there was evidence for stars at least fifty times more massive than the sun. Because of their great mass and, consequently, the great pressure in their cores, they burn their fuel rapidly and survive only ten million or twenty million years. Countless numbers of them must have burned out in the many billions of years since stars were first formed. What has happened to them?

"A star of large mass," wrote young Chandra, "cannot pass into the white-dwarf stage, and one is left speculating on other possibilities."

He modestly refrained from specifying such possibilities, but his suggestion of continuing collapse was not warmly received. The *Astrophysical Journal* (of which Chandra, for almost two subsequent decades, was the meticulous editor) rejected his initial paper, and when it was published elsewhere, those whom he later described as "the stalwarts of the day"—men like Sir Arthur Eddington and Edward A. Milne, professor of mathematics at Oxford, a leading cosmologist (and Chandra's friend), found his idea preposterous. "To me," Milne wrote Chandra, "it is clear that matter cannot behave the way you predict." Eddington commented:

> Chandrasekhar shows that a star of mass greater than a certain limit . . . has to go on radiating and radiating and contracting and contracting until, I suppose, it gets down to a few kilometers' radius when gravity becomes strong enough to hold the radiation and the star can at last find peace.

Had Eddington stopped there, as Chandra later pointed out, "we should now be giving him credit for having been the first to predict the occurrence of black holes. . . ." Eddington, however, went on to describe this as "almost a *reductio ad absurdum*"—a line of reasoning carried to the point of absurdity: "I think that there should be a law of Nature to prevent the star from behaving in this absurd way."

When Chandra presented his arguments he was not necessarily thinking in terms of total collapse. Perhaps, he wrote later, a collapsing superstar throws off enough material to reduce itself to a stable size. In any case, it was another decade before scientists pursuing problems far removed from the puzzle of unbridled collapse showed that white dwarfs are not the final stop on the road to oblivion—that objects of even greater density are possible. They were working on such questions as what suddenly causes some stars to increase their brilliance one hundred billionfold.

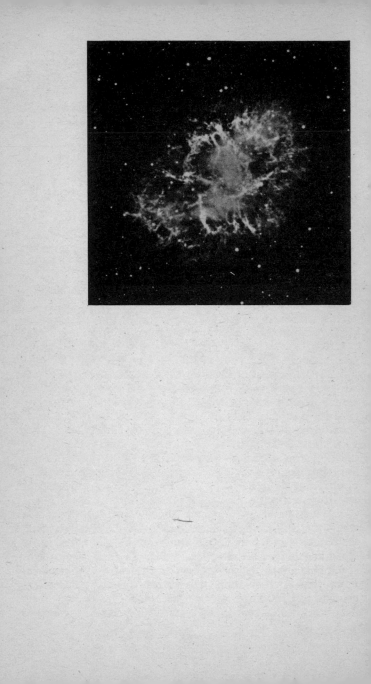

3

Brighter Than
a Hundred Billion Stars

"I make my kowtow," wrote the astronomer royal of Imperial China, Yang Wei-t'e, in 1054. "I observed the phenomenon of a guest star. Its color was slightly iridescent. Following an order of the Emperor, I respectfully make the prediction that the guest star does not disturb Aldebaran [a bright star nearby in the constellation Taurus]; this indicates that . . . the country will gain great power. I beg to store this prediction in the Department of Historiography."

The "guest star" that appeared seemingly from nowhere was so bright that for months it could be seen in broad daylight. Within a year or two, however, it vanished again. This, we know today, was a "supernova"—a cataclysmic explosion in which a star suddenly flares up until it may be brighter than the combined light of all the one hundred billion other stars of the galaxy, or star system, to which it belongs. The galaxy that is the home of the sun and all its planets, including the earth, is partially visible as the Milky Way, a luminous band across the sky. Although its billions upon billions of stars and its dust clouds are arranged in a great flattened spiral, we are too deep within the pattern to see it.

When telescopes today are aimed at the spot in Taurus where the Chinese and Japanese recorded that "guest star," the remnants of the explosion are visible as a torn, luminous cloud known as the Crab Nebula. It looks much like a high-speed photograph that has caught an explosion in midcourse. From various observations it is known that the cloud is expanding at 1,100 kilometers per second and at present is six light-years wide—that is, so wide it takes light six years to

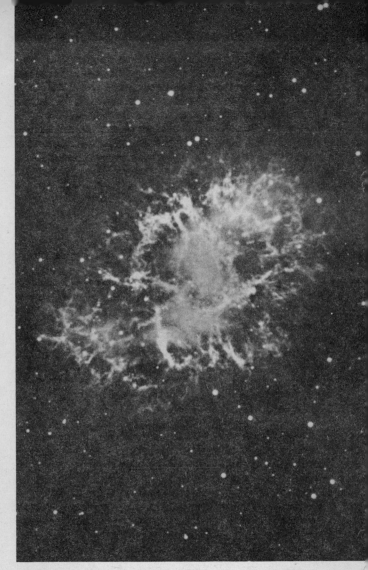

The Crab Nebula, whose explosive origin in the supernova observed in A.D. 1054 is clearly shown in its appearance. This twenty-minute exposure was made in red light on February 27, 1976, by William van Altena of Yale University with the four-meter telescope of the Kitt Peak National Observatory.

Fritz Zwicky

cross it, traveling at 300,000 kilometers a second. The supernova of 1054 actually took place some six thousand years earlier for, while it lies within the Milky Way Galaxy, it is six thousand light-years away.

Supernovas that occur in the visible parts of our own galaxy are seen only once in every few centuries. The most complete record is that of the Chinese, who have recorded a half dozen or so in the past two thousand years. It is unlikely that they missed any because over that period they recorded without exception every reappearance of Halley's Comet, which returns roughly every seventy-six years. They also recognized the difference between such regular revisits and the "guest stars" that appeared only once and with great brilliance.

The most recent supernova observations within our galaxy were by two of the great Renaissance students of the heavens: Tycho Brahe and Johannes Kepler. It was Brahe's very precise observations of planetary movements that enabled Kepler to guess the nature of their orbits around the sun. In 1572 Brahe recorded the sudden appearance of a very bright star in the constellation Cassiopeia. He called it a "nova" or new star. Today, because a number of stars flare up to a moderate degree—some of them previously invisible—one that does so in the far more spectacular manner is called a supernova. Kepler described one in 1604, and ever since that time, astronomers have impatiently awaited another such spectacle, hoping during the hours of its greatest brilliance to capture enough data, using the most modern devices, to decipher in detail what happens in such a cataclysm—now known to be the death throes of a large star.

Supernovas presumably bathe surrounding space with intense radiation, and it would be unfortunate if such an explosion occured near an inhabited world, as it would probably be fatal to the life on that planet.

In the 1930s Fritz Zwicky of the California Institute of Technology became impatient with waiting for a local supernova. Zwicky, a man with a rather lionlike head who was born in Bulgaria of Norwegian parents and educated in Switzerland, was often impatient. He was regarded by some as "tempestuous," but also, in addition to major contributions in astronomy, as a pioneer in the development of jet engines.

Zwicky realized that if, in the search for an explanation of supernovas, one waited until there was such an occurrence in this galaxy, generations might come and go before this happened. He therefore began looking at other galaxies resembling our own Milky Way on the theory that, if they occur on the average once every two centuries in any one galaxy, they should appear once every two years in a population of one hundred galaxies. Today scanners of the skies with the new powerful telescopes are seeing a dozen supernovas every year, although they are too dim for detailed study.

Zwicky and his German colleague Walter Baade, who worked at the Mount Wilson Observatory overlooking Cal Tech, analyzed the records of a supernova seen in the Spiral Nebula of Andromeda in 1885. This "nebula" is actually the nearest spiral galazy resembling our own. They calculated that in twenty-five days that one star radiated as much light as does the sun in ten million years. Yet it is no longer visible. Far closer was the supernova of 1572, which occurred in our own galaxy. Yet, said Zwicky and Baade, "repeated attempts" to associate it with even a faint star near its position in Cassiopeia "have so far not been very convincing." What had happened to it?

In the first place, the two men reasoned, when such an event occurs, a large part of the star's mass must be converted to energy. To explain what remained they turned to some current speculation—notably by two Russians—on what happens when atoms are so severely squeezed that their "electron pressure" is overwhelmed. One of them, George Gamow, was an originator of the now widely held view that the universe was born in a primordial explosion or "Big Bang." In 1933 he left the University of Leningrad for George Washington University in Washington, D.C. Gamow described what happens in the cores of stars as neutron decay running backward.

As noted earlier, neutrons, when outside the atomic nucleus, decay into protons by shooting off an electron (and an elusive antineutrino). Half a given number of neutrons decay in this manner every thousand seconds. Inside a stable nucleus, on the other hand, they do not decay. If the pressure on a gas containing protons and electrons is sufficient, Gamow said, the reverse process will occur. Electrons will be forced to mate with protons, forming neutrons. Therefore, he added, the cores of very massive stars must consist of "a gas of neutrons" so dense that it becomes "analogous to the conditions inside an atomic nucleus."

He cited the proposal of his countryman, Lev Landau, that atoms are crushed within the cores of stars to form—as Landau put it in 1931—"one gigantic nucleus." Landau's career as a Nobel Prize winner and one of Russia's most brilliant theorists was cut short in 1962 when a head injury in an automobile accident left him temporarily blind, deaf, and speechless. He recovered to a considerable extent but died six years later.

In a paper submitted to a Soviet journal in 1931 (a year after Chandra's first white-dwarf analysis) Landau, then in Zurich, argued that for stars with more than 1.5 times the mass of the sun (that, for simplicity of analysis, were not generating energy) the pressure on their cores would be so great that no forces envisioned in contemporary atomic theory coud resist it. For such situations, he wrote, "there exists in the quantum theory no cause preventing the system from col-

lapsing to a point." He noted that electrical repulsion between particles of similar charge, such as negatively charged
electrons or positively charged protons, becomes inconsequential under such circumstances.

Since objects with masses much larger than that of the
sun "exist quietly as stars" and do not show "such ridiculous
tendencies" as collapsing to a single point, Landau said, the
rules governing atoms must break down in the stellar cores
and all stars must possess "pathological regions" of highly
condensed matter. He later saw such cores as formed of
closely packed neutrons and argued that stars, in fact, derive
their energy through the gradual collapse of atoms to form a
neutron core of slowly increasing size. To generate the energy radiated by the sun in the past two billion years only 2 per
cent of the sun need have collapsed this way, Landau said.

Zwicky and Baade proposed that a similar, tightly
packed ball of neutrons is the residue of a supernova: The
burned-out star collapses, crushing its protons and electrons
into neutrons; the cataclysmic infall of material releases
enough gravitational energy to produce all the manifestations
of such an event; and a superdense residue of neutrons is the
inevitable end product.

"With all reserve," said Zwicky and Baade, "we advance the view that supernovae represent the transitions from
ordinary stars into neutron stars, which in their final stages
consist of extremely closely packed neutrons."

Whereas the material forming a white dwarf weighs
about ten tons per thimbleful, a similar volume in a neutron
star would weigh one hundred million tons. The star would
be less than twenty kilometers in diameter and very difficult
to see across the vast distance between the earth and even the
nearest stars. And so there seemed no way to confirm the proposal. Yet it was tantalizing to think of an object so dense
that (as described in the next chapter) the gravity it exerted,
in terms of relativity theory, would radically warp space and
slow time. As Zwicky and Baade put it, "the fascinating
problem now presents itself of investigating how certain well-
known physical processes . . . will be modified when they
take place inside of highly collapsed stars in which the very
properties of time and space are drastically altered."

The probability that neutron stars would never be observed did not discourage further speculation. Scientists like to play with ideas, even when they lie beyond obvious means of confirmation. That is how some of the most basic advances have come about, as with Einstein's youthful "thought experiments" in which he tried to imagine how things would appear if he were riding on a beam of light.

It was such thinking that led Einstein to his first great achievement—the "special" theory of relativity, first published in 1905. It predicted the strange effects of motion at high velocity, such as the slowing of time, and it led to his formulation, two years later, of the relationship between mass and energy, most dramatically demonstrated by atomic bombs.

But the theory was "special" in that it did not include a "relativistic" treatment of gravity—one that would be valid regardless of the observer's motion or location. It was in 1907 that, as Einstein put it, "there came to me the happiest thought of my life." He saw a way to incorporate gravity and make his theory "general" instead of "special." The resulting "general theory of relativity," published in 1915, set the stage for others to calculate more realistically what happens when stars collapse and how strong gravitational fields might cause objects to disappear.

4
The Warp of
Space and Time

To describe Einstein's general theory of relativity—the heart of the modern black-hole concept—in a way that satisfies physicists it must be done in mathematical terms. Nevertheless, those elements of it that determine the nature of black holes can be sketched more simply. Of particular importance are the effects of gravitational fields on space and time.

Prior to World War I Einstein came to what was then an extraordinary conclusion: Gravity, particularly when very strong, warps space and slows time. Unlike the concept of an electromagnetic field operating in three-dimensional space, he saw gravity in four-dimensional terms involving time as well as space. His first steps in that direction were relatively straightforward and easy to comprehend. They were based on arguments concerning the equivalance of gravity and acceleration.

To understand such equivalence imagine that you are in a fully equipped laboratory aboard a very large spaceship. You awaken after a long sleep and realize that one of two situations may exist: either the spaceship is still sitting on the launch pad and what keeps you from floating around is the earth's gravity, or the ship has been launched and is accelerating upward at a rate that mimics the force of gravity to which we are all accustomed (that which causes a falling body to increase its velocity every second by thirty-five kilometers an hour). There are no windows enabling you to settle the matter by looking out.

You throw a ball gently across the laboratory and its flight curves down until it hits the floor, but that does not answer the question. The curvature could be due to gravity, or it could occur because, during the ball's flight, the upward motion of the floor (and all the rest of the spaceship) accelerated. As Einstein pointed out, there is no experiment that could be done in the laboratory to tell the difference between gravity and acceleration. They are equivalent and, in essence, the same. Stated another way, the inertial mass of a body (its resistance to acceleration) is proportional to its response to gravity (its weight). The effect on light of gravity and acceleration therefore should be identical.

To demonstrate the effect of acceleration—and hence of gravity—our hypothetical laboratory, since light moves so fast, must be accelerating "upward" at a very high rate—far higher, in fact, than a human body could withstand. But our imaginations can witness the test. This time, instead of a ball being thrown across the lab, a pulse of light is projected across it, passing through a succession of fluorescent glass plates. As the pulse goes through each plate it leaves a glowing spot. It is easy enough to see that, if the laboratory is accelerating upward, those spots will not form a straight line but will curve down toward the floor, like the path of the ball. The reason is that as the pulse of light travels from plate to plate the upward velocity of the lab increases enough so that the light pulse is, to some extent, "left behind." The light has been "bent" by acceleration and should be bent by gravity, particularly if the latter is very strong.

Now, in this accelerating laboratory, let us conduct an experiment using a clock that emits a little flash of light every second. In this case the laboratory, in the direction of acceleration, is very long (to make the effect more obvious). The clock is placed at the rear end and our rugged space traveler watches it from the front end. If the movement of the spacecraft were uniform, the distance light had to travel between clock and observer would have remained uniform and the once-a-second flashes would have matched the second hand on the observer's wristwatch.

But if the laboratory is accelerating, the distance to be traveled between the clock, at the rear of the lab, and the observer, at the front end, keeps increasing. The light pulses, having to travel farther and farther, on arrival will lag behind the observer's second hand.

Looking at a clock in a strong gravitational field is equivalent to looking at the clock at the far end of an accelerating laboratory. It will seem to run slow, and anything else that is time-dependent will also "run slow." This includes the oscillations of light waves and radio waves. Their vibrational frequency will be slowed and their wavelengths will lengthen. Therefore the characteristic wavelengths of light emitted by atoms in a massive star—one with strong gravity—will be somewhat reddened. This is the "gravitational red shift" that was to play an important role in the black-hole concept.

In 1911 Einstein saw a way to test the light-bending effect of gravity. Although some stars are visible through telescopes in daylight, those in directions close to that of the sun are blotted out by the sun's glare. They can, however, be observed when the moon cuts off the sun's rays during a total solar eclipse. If light rays from stars beyond the sun were bent when passing close to the sun, where the latter's gravitational effect is strong, the stars should seem out of place. Einstein calculated that if the starlight just grazed the sun the apparent displacement would be 0.83 second of arc—the angular width of a twenty-five-cent piece four miles away. A reworking of Einstein's calculations has shown that he should have said 0.87 second, but as the physicist Banesh Hoffmann has gently remarked, "arithmetic was never one of Einstein's strong points."

Einstein wondered whether astronomers knew of some trick whereby they could conduct the test without having to wait for an eclipse. In 1913 he wrote to George Ellery Hale, founder and head of the Mount Wilson Observatory in California, asking if that might be possible. Hale said no.

However, a solar eclipse was to occur in Russia the next year, and the German astronomer Erwin Finlay-Freundlich planned to seize the opportunity to make the critical test. As pointed out by Hoffmann in his biography of Einstein (done in collaboration with Einstein's long-time secretary Helen Dukas), it was fortunate for the discoverer of relativity that the outbreak of World War I prevented the observation, for Einstein was wrong. As he recognized a few years later, the effect would be twice what he had first predicted—1.7 seconds of arc.

What Einstein's original calculation had overlooked was that there are two gravitational effects arising from his general theory—one a slowing of time and the other a curvature of space. If, in grazing the sun, light passed through strong gravity, its velocity (being time-dependent) would be slowed from the viewpoint of an external observer. When light waves emerge obliquely from a region in which they have been slowed, as in passing through water, they are bent—an effect that makes a partially submerged oar or stick seem bent.

Einstein's 1911 calculation, based on the equivalence principle, had taken into account only one effect. Later, when his general theory was fully developed, he combined all the effects of gravity on both space and time and found that the bending would be double what he first thought.

By now Einstein's homeland—Germany—was at war, but an account of his theory and his prediction regarding a solar eclipse crossed the battle lines and reached Sir Arthur Eddington in England. Eddington was greatly excited and soon became the theory's most eloquent proponent. He had also mastered its subtleties. As he was leaving a meeting at which the eclipse test was discussed, another theorist (Ludwig Silberstein) remarked to him: "Professor Eddington, you must be one of three persons in the world who understands general relativity." Eddington did not respond for a few moments, and Silberstein added: "Don't be modest, Eddington," whereupon the latter replied: "On the contrary, I am trying to think who the third person is!"

Eddington worked hard to persuade his countrymen to ignore the fact that the theory had originated in a land with which Britain was locked in mortal combat. There was soon to be a rare opportunity to test the prediction—one that the world of science, which ideally knows no national boundaries, could ill afford to let pass. As Eddington later pointed out, "the most favorable day of the year for weighing light is May 29." On that date the earth is so placed in its orbit that the region beyond the sun is exceptionally rich in stars—the cluster known as the Hyades or "Rain Bringers" in the constellation Taurus, the Bull.

"Now if this problem had been put forward at some other period of history," wrote Eddington, "it might have been necessary to wait some thousands of years for a total eclipse of the sun to happen on the lucky date." But, he added, "by strange good fortune" such an eclipse was to occur on May 29, 1919, its zone of totality traversing the Atlantic from Africa to Brazil.

In 1917, although Britain was suffering through one of the grimmest phases of the war and U-boars might sink ships en route to the sites, plans were laid for two expeditions, one from the Royal Greenwich Observatory and the other from the University of Cambridge, where Eddington was Plumian professor of astronomy. The observations would be made from the island of Principe off Africa in the Gulf of Guinea, and across the Atlantic at Sobral in Brazil.

Those taking part in the expeditions discussed the possible outcomes with Sir Frank Dyson, the astronomer royal, who was co-ordinating the plans. Perhaps, it was said, Einstein's prediction would be confirmed. Or it seemed possible that his original guess of half that amount would prove correct. There might be no deflection at all, conforming to conventional physics. Finally, asked E. T. Cottingham, who was to go with Eddington to Principe, "What will it mean if we get double the deflection?" In that case, said Dyson, "Eddington will go mad, and you will have to come home alone."

By the time of the eclipse the war had ended. On Principe the day began with rain and not until the eclipse had started did the observers see the sun at all. "I did not see the eclipse," wrote Eddington in his diary, "being too busy changing plates, except for one glance to make sure it had begun and another halfway through to see how much cloud there was. We took sixteen photographs. . . . The last six photographs show a few images, which I hope will give us what we need. . . ." Eddington began immediate attempts to measure the star positions, although more careful measurements would be possible when they got home. "Three days after the eclipse, as the last lines of the calculation were reached," Eddington recorded, "I knew that Einstein's theory had stood the test and the new outlook of scientific thought must prevail. Cottingham did not have to go home alone."

The Sobral expedition had better luck on the weather, and their results provided added support. The deflections calculated by the two expeditions were 1.98 and 1.61 seconds, bracketing Einstein's prediction of 1.70. As Sir James Jeans put it, the expeditions had returned "bringing back news which changed, and that irrevocably, the astronomer's concept of gravitation and the ordinary man's conception of the nature of the universe in which he lives."

While the light-bending of general relativity was established by the 1919 eclipse—and by a number of later tests—the influence on time, particularly the "twin paradox," based, perhaps, on a mixture of effects from general and special relativity, has been harder for people to accept. It predicted that a twin who went on a long space journey at high velocity would return substantially younger than his brother.

In 1911 Einstein was invited by the Natural Science Society of Zurich to explain his theories at one of its meetings. "The most droll consequences arise," he told the Swiss scientists, when a clock is sent on a long trip, traveling almost at the speed of light, and is then brought back at similar velocity.

It turns out [he said] that during the entire journey the hands of this clock have hardly moved, whereas the hands of a clock that remained stationary at the starting point throughout that time have changed very substantially. One must add that what applies to this clock, which we offer as a simple representation of all physical phenomena, also applies to a closed physical system of any other kind. If, for example, we enclose a living organism in a box and send it on the same trip, out and back, as previously we did for the clock, it could come about that this organism, after an arbitrarily long flight, will be arbitrarily little changed upon return to its orginal location, whereas organisms having entirely equivalent properties but which remained at rest at the original site will long since have made way for new generations. If the speed of the moving body approached the speed of light, the long duration of the journey will have been, for that moving organism, but a moment! That is an incontrovertible consequence of the basic principles we have postulated as forced upon us by experience.

He presented the effect as due entirely to uniform motion, rather than acceleration. Gerald Holton of Harvard, an authority on the origins of relativity, who drew my attention to this early exposition of the twin paradox by Einstein, pointed out: "This was, after all, still 1911, and he was quite aware of the need for much more work before general relativity could be taken seriously." Some theorists today, in fact, regard the effect as entirely a by-product of the original "special" theory in which the speed of light plays a central role. Others relate it to the general theory, dealing with gravity and acceleration, since any such traveler would be subject to large accelerations at the start, turn-around, and end of his trip. In his presentation to the Swiss group, however, Einstein said he did not include acceleration in his calculations, saying it was not clear how to do so.

Doubters of the twin paradox have argued that it simply involves Doppler effects that cancel out on the return journey. If, for example, one brother sets off at four fifths of the speed of light with a clock that emits a brilliant flash once an hour, and his twin stays at home with a similar clock, then, because the two are separating so fast, each will see the flashes of the other's clock every three hours instead of once an hour. Just as the pitch of a horn is reduced (its waves lengthened) by such motion, so will the rate of the light flashes.

On the return journey of the traveling twin the reverse effect will occur. Each brother will see three flashes an hour. The speed-up will cancel the earlier slowdown and on return both clocks will read the same.

The fallacy of this argument was demonstrated in 1957 by Sir Charles Darwin, grandson of the father of evolution theory. In a letter to *Nature* he pointed out that if a journey, such as the one described above, carried one twin to a point four light-years away, traveling at four fifths of the speed of light it would take him five years to get there and five years to return. As did Einstein, he purposely omitted the effects of acceleration, prescribing that observations would be made only when the spaceship was up to speed.

[71]

As soon as the traveler reverses direction, said Sir Charles, he will observe the increase in the rate of flashes from his stay-at-home brother, but this will not be true of the latter. Since his traveling twin reverses course four light-years away, the speeded-up flashes will not begin reaching earth until four years later. When the space voyager returns his brother will only have seen one year of fast flashes and the traveler will be correspondingly younger.

No one has come up with a convincingly practical way to accelerate a spacecraft for long periods, but it has been calculated that, if such a vehicle was continuously accelerated at a rate comparable to that of a falling body on earth, this would simulate the effect of gravity for those on board, leading to a comfortable journey. After a year the craft, relative to the earth, would be approaching the speed of light and from then on could not increase its speed substantially. Instead, from the point of view of those at home, strange "relativistic" effects would come into play. Time on board will seem to be virtually at a halt. After a round trip that, for those on board, lasted twenty years, taking into account all the effects of relativity, they would find that 270 years had elapsed on earth. If the journey lasted sixty years for the travelers, the elapsed time on earth would be five million years!

To achieve prolonged accelerations schemes have been proposed whereby, for example, a spacecraft would scoop up hydrogen en route as fuel, but no such idea has seemed very feasible. Fuel limitations are illustrated by the two Viking missions, launched toward Mars in 1975. The massive, multi-stage combination of rockets that sent them on their way burned, in each case, for a total of twenty minutes—a far cry from accelerations measured in years. Virtually all the rest of the 220-million-mile journey to Mars was coasting flight.

Typical of skepticism about the twin paradox was an article published in *Physics Today* in 1971 by Mendel Sachs, a professor in the physics department at the State University of New York in Buffalo. He conceded that his was a minority view, but he argued that the effects of general relativity were being misinterpreted. The traveling twin, he said, would come home no older than his brother.

It was in that year that a way to test the paradox was recognized by Joseph C. Hafele, a physicist from Washington

University in St. Louis. He and Richard E. Keating of the United States Naval Observatory's Time Service in Washington, D.C., were able to demonstrate the effect using atomic clocks and scheduled jet airliners.

The Naval Observatory, chief timekeeper for the United States, is equipped with a number of portable atomic clocks that depend for their extraordinary accuracy on oscillations that occur in atoms of cesium 133 as they fluctuate between two energy states. The atoms, when stimulated at the proper frequency, emit microwaves whose highly stable frequency is 9,192,631,770 cycles per second. The clocks, using this as their rate control, do not vary by much more than a billionth of a second per day.

This, the two scientists realized, was accurate enough to show the very slight departures from normal time to be expected in their experiments, particularly if four such clocks were carried and their rates averaged.

The two men flew around the world twice, first to the east and then to the west, riding commercial jets that normally fly at elevations of several miles. Relativity was expected to influence the clocks in two ways. One was the time-slowing effect to be expected from a high-velocity round trip, popularly attributed to special relativity. While it would become large when the speed of light is approached, at jet-plane speeds it would be detectable only by atomic clocks.

The other factor working on the passage of time would be from general relativity. According to that theory, clocks on the top floor of a house should run a tiny bit faster than those on the bottom floor, because they are farther from the center of the earth and therefore in a slightly weaker gravitational field. To the satisfaction of most physicists this seemingly preposterous effect had already been demonstrated in the seventy-foot tower of the Jefferson Physical Laboratory at Harvard University. The experiments had been conducted there in 1959 and 1960 by Robert V. Pound, a Harvard physics professor, and Glen A. Rebka, Jr., a graduate student. They used an ingenious experimental setup dependent on the newly discovered "Mössbauer effect," which enabled them to detect extremely slight variations in wavelengths (and, therefore, oscillation frequencies) of emitted light waves (or other forms of electromagnetic radiation). They showed, for example, that gamma-ray waves emitted by nuclei of iron 57 were longer (and the oscillation rates slower) when the nuclei were at the bottom of the tower than when at the top. This was the "gravitational red shift" predicted by the theory and also represented a slowing of time within the slightly stronger gravitational field at the foot of the tower. (A related experiment was taken as evidence for a slowing of time due to acceleration.)

These experiments, however, had not convinced the skeptics. The difference between gravity on the ground and at altitudes flown by jetliners would be considerably greater than that between the top and bottom of Harvard's Jefferson Physical Laboratory. Since the time-slowing effect of gravity at jet cruising altitudes would be less than for the clocks left behind at the observatory, clocks on round-the-world planes should run a little faster.

The eastward trip took 65.4 hours, of which 41.2 hours were spent in flight. The westward trip lasted 80.3 hours, of which 48.6 were in the air. The effects would depend in part on the flight path, speed, and altitude. The plane captains provided the necessary information and the eastward trip was divided into 125 intervals in each of which conditions were relatively uniform and subject to a single mathematical analysis. The westward trip was likewise divided, into 108 intervals.

From all of these data the two scientists calculated what effect the two relativity theories should have on the clock rates. Because of general relativity the clocks should have gained 144 nanoseconds (billionths of a second) flying east and have gained 179 nanoseconds flying west. The predictions were not identical because flights in the two directions were, on the average, at different altitudes and therefore in slightly different gravitational fields.

The predicted effect of special relativity on the eastward flight was a loss of 184 nanoseconds. This took into account the fact that the motion of the airborne clock to the east was much faster than that of the comparison clocks in Washington, which with the earth's rotation were themselves moving east at about 1,275 kilometers per hour. The total speed of the flying clocks was the eastward flight of the plane plus the eastward motion of the earth's surface and its atmosphere due to rotation.

On the other hand, when the clocks were carried west, the effect was assumed to have been opposite. Now they were flying contrary to the earth's spin. Jet airliners, unless they are supersonic, cannot in most instances fly as fast as the surface motion of the earth due to the planet's spin (motion we do not, of course, feel, as everything around us, including the air, is moving east together). Thus when one flies from New York to Los Angeles through the east-moving atmosphere, although flying west one still moves east relative to some "stationary" reference point, such as the earth's center. If the flight speed fully canceled the rotation speed (as on westward transatlantic flights of the supersonic Concorde), the local time of arrival in Los Angeles would be identical to, or even earlier than the time of departure from New York, but that is not the case. The flying clocks therefore still moved east, but more slowly than the stay-at-home clocks, and as a result they should have gained 96 nanoseconds (these estimates all being subject to error margins of 10 to 18 nanoseconds).

When the effects of the two forms of relativity were combined, the prediction was for a net time loss of 40 nanoseconds flying east around the world and a net gain of 275 nanoseconds flying west.

The actual gains or losses in time recorded by the clocks in nanoseconds were:

CLOCK NUMBER	EASTWARD	WESTWARD
120	—57	277
361	—74	284
408	—55	266
447	—51	266

The clocks were thus much in accord with one another and with the predictions. The experimenters described the similarity as "very satisfactory." There seems, they added, "little basis for further arguments about whether clocks will indicate the same time after a round trip, for we find that they do not."

What they did not mention was that, because of the combined effects of slowed time on their eastward journey and its faster rate when flying west, they had come home roughly 233 nanoseconds older than if they had stayed at home. One could claim this was a small but finite price to pay for so clearly demonstrating the time-warping effects of relativity, but in fact they did not miss out on anything. Not only their clocks but everything else in their lives moved faster. Had the effect been far more radical they would have enjoyed (or hated) every little shrunken second of their lives just as much as a full-length second, for to them the passage of time would have seemed normal.

Since then a new series of experiments has been carried out in a joint project of the Naval Observatory and Carroll Alley of the University of Maryland. In this case cesium clocks, instead of circling the world, were carried by an aircraft circling at a succession of altitudes above the Patuxent Naval Air Test Center on Chesapeake Bay. Similar clocks remained in a van on the airfield. After each of five fifteen-hour flights, completed in the spring of 1976, the plane was parked alongside the van to facilitate direct comparison of the two sets of clocks, and it was shown that those on the ground had run slower. Furthermore, by means of extremely short ground-to-air laser pulses, it was possible to compare clock rates during the flights.

As the plane burned up its heavy fuel load it climbed from 25,000 feet (7,600 meters) to 30,000 feet and then 35,000 feet. With each increase in altitude its clocks increased their rates relative to those in the van in a slight but observable manner. In view of such tests no doubt remains that gravity slows time and bends light—or, more properly, warps space.

5

The Black Hole Idea Born

The birth of the black-hole concept, in modern form with all of its strange properties attributable to relativity, is usually credited to J. Robert Oppenheimer, "father" of the atomic bomb. The first major step in that direction, however, was taken by a German astronomer, Karl Schwarzschild, as he lay stricken with a fatal illness in the winter of 1915–16.

One of Germany's most brilliant astronomers and theorists, he was also determinedly patriotic. As head of the Potsdam Astrophysical Observatory and a man already in his forties, he could have escaped military service, but he volunteered for active duty on the Russian Front, where he contracted pemphigus, a rare, insidious, and (then, at least) invariably fatal disease. A few months earlier Einstein had published the essential elements of his "general" theory, and one of the three classic papers that Schwarzschild wrote on his deathbed dealt with special implications of that theory.

To explore in the simplest possible way what would happen when gravity became extremely strong, he treated gravitational fields as emanating from bodies with no volume whatsoever. He assumed that all the mass of an object, such as the earth or sun, was concentrated into an infinitesimal point.

Using Einstein's equations as a guide, he found that, as one approaches such a point, the effects of relativity begin to soar until space curvature closes in on itself and the red shift becomes infinite—light waves and other waves become infinitely long. Such a region would be completely cut off from the outside. His surprising discovery was that these effects occur before one reaches the hypothetical focus of gravitational attraction. If a mass comparable to that of the earth were concentrated into a point, the radius at which these effects occurred would be roughly one centimeter. It came to be known as the gravitational radius or Schwarzschild radius.

Schwarzschild sent his paper to Einstein who, on January 9, 1916, replied:

> My esteemed Colleague!
>
> I have read your paper with the greatest interest. I had not expected that the exact solution of the problem could be formulated so simply. The analytical treatment of the problem appears to me splendid. Next Thursday I will present the work, with several explanatory words, to the Academy.

After an extensive discourse on other theoretical questions he ended:

> Wishing you the best
> Your devoted
> A. Einstein

Four months later Schwarzschild was dead.

It was ultimately recognized that the Schwarzschild radius was the gate to the black hole. If, at that boundary, gravitational effects became infinite, nothing could escape. Even light would be trapped, if only because its wavelengths, stretched by the gravity, would become infinitely long.

Nevertheless, at the time the concept was considered an academic exercise, a theoretical curiosity. Sir Arthur Eddington discussed it in his classic 1926 work *The Internal Constitution of Stars*. He pointed out that the star Betelgeuse (one of the four corner stars of Orion) exerts very potent gravity, having ten to one hundred times the mass of the sun. But its mass is not sufficiently concentrated to come near the intensity needed to make it invisible. The star is fifty million times more voluminous than the sun—as big around as the orbit of the earth. He conceded, however, that if it was as dense as the sun, then indeed strange things would happen:

Firstly, the force of gravitation would be so great that light would be unable to escape from it, the rays falling back to the star like a stone on the earth. Secondly, the red-shift of the spectral lines would be so great that the spectrum would be shifted out of existence. Thirdly, the mass could produce so much curvature of the space-time metric that space would close up around the star, leaving us outside (i.e., nowhere).

It was a few years later that Eddington, commenting on the idea that some stars may go on collapsing indefinitely, said there should be a law of nature "to prevent the star from behaving in this absurd way."

Eddington noted that as early as the end of the eighteenth century the great French mathematician and astronomer Pierre Simon, Marquis de Laplace, had recognized the possibility that, if stars were massive enough, they might become invisible. Their gravity would be so strong, "corpuscles" of light could not escape them. Laplace knew that light has a finite—rather than infinite—speed. This had been demonstrated in the previous century by Ole Rømer, a Danish astronomer who found that when Jupiter was on the side of the sun opposite the earth, and therefore at maximum distance, its moons were always "behind schedule" in their orbits around that planet, compared to their timetables when Jupiter was closest. He recognized that they seemed late because of the extra distance the light had to travel, and from this he calculated the speed of light to be 225,000 kilometers per second. It is now known to be 299,792.5 kilometers a second, but that was remarkably accurate considering the manner of measurement.

Laplace assumed that light, like everything else with which he was familiar, is slowed by gravity. The strength of the earth's gravitational field is such that an object must be traveling at 40,000 kilometers an hour as it departs to escape. Gravity slows but does not stop its ascent. At any lower initial velocity the object slows and finally falls back to earth.

The escape velocity on the moon, with its smaller mass, is 8,570 kilometers per hour. Those who watched the astronauts lope across the lunar surface were made vividly aware

Albert Einstein in his later years.
(The New York Times)

Karl Schwarzschild

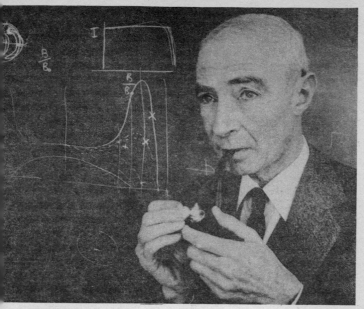

J. Robert Oppenheimer
(Wide World)

of its far weaker gravity, and it took much less rocket power to get them off the moon than it did to leave the earth. On the other hand, escaping Jupiter, by far the most massive of planets, would require almost 220,000 kilometers per hour. The escape velocity of some stars, Laplace argued, may exceed the speed of light.

He was aware that some heavenly bodies are very different from our sun—much larger, much brighter, much less stable. "Several stars," he wrote in 1798, "display in their color and their brightness very remarkable periodic variations. There are others that have appeared all of a sudden, and that have disappeared after having, for some time, emitted a brilliant light. What prodigious changes must have taken place on the surface of these great bodies, to be so evident at the distance that separates us from them? How much they must surpass those that we observe on the surface of the sun, and convince us that nature is far from being always and everywhere the same?''

Little did he realize the scope of the ''prodigious changes'' that are, in fact, occurring far out in space. ''All these bodies that have become invisible,'' he continued (he called them, in French, *corps obscurs*), ''are in the position where they were observed, since they did not change it at all during their appearance. There exist, therefore, in heavenly space invisible bodies as large, and perhaps in as great number, as the stars. A luminous star of the same density as the earth, and whose diameter was two hundred and fifty times greater than that of the sun, would not, because of its attraction, allow any of its rays to arrive at us; it is therefore possible that the largest luminous bodies of the universe may, through this cause, be invisible.'' Since Laplace believed massive stars slow down the light radiating from them, he thought this must affect the apparent position of those not large and dense enough to be invisible. We see them where they were longer ago than neighboring stars that are smaller.

He omitted his discussion of *corps obscurs* from later editions of his classic book on astronomy, *Exposition du Système du Monde*. Perhaps he had doubts about it. By a remarkable coincidence the gravitational field that Laplace calculated was necessary to make an object invisible was close to that estimated by Schwarzschild, even though they attacked the problem from basically different points of view. Laplace worked from the misconception that light ''corpuscles'' can be slowed by gravity in the same manner as material objects, and he thought in terms of stars as big as the circumference of the earth's orbit around the sun. Schwarzschild applied relativity theory to objects of infinitesimal size.

A century after Laplace made his proposal, in the 1880s, Albert A. Michelson and Edward W. Morley at the Case School of Applied Science (now Case Western Reserve University) in Cleveland, Ohio, showed that measurements of the speed of light through a vacuum always come out the same, whether the light source is stationary or in rapid motion toward or away from the observer. The finding seemed very odd until Einstein showed how it fitted into his special theory of relativity. His general theory then showed that gravity only affects the speed of light by slowing time in a region through which light may have passed (as in the eclipse test), and that gravity stretches the waves of light and bends their path, making it possible for stars to become invisible for reasons quite different from those envisioned by Laplace.

It was not the abstract calculations of Schwarzschild or the discussions of Eddington that finally led J. Robert Oppenheimer to produce the first comprehensive description of a black hole. It was the debate about neutron cores and the collapse of dying stars that set him on the track, notably Lev Landau's discussion of the possibility—which the Russian quickly dismissed as "ridiculous"—that any star much more massive than the sun, once its internal energy production ceased, "will have a tendency to collapse to a point."

The name Oppenheimer today conjures up memories of a grim period when the man who, during World War II, had headed the atomic-bomb project at Los Alamos, New Mexico, became a victim of panicky Cold War fears and, it seems, a certain amount of personal spite, losing the security clearance that had enabled him to play a leading role in determining weapons policy.

In 1929, when "Oppie"—as he came to be known to his colleagues—began working at Cal Tech and Berkeley, he was a lean young man of twenty-five who had graduated *summa cum laude* from Harvard after completing the four-year course in three years. He kept his mind busy on all sorts of projects—for example, learning Sanskrit, the ancient language of India, so he could read its literature in the original. He became a magnet for brilliant students, and it was with several of them that he began exploring what rules might govern the behavior of tightly packed neutrons. One of them was George M. Volkoff, a graduate student who had been born in Moscow in 1914 and, after the upheavals of the Russian Revolution, arrived in Canada at the age of ten. He returned to Harbin, Manchuria, for his intermediate education, then came back to the University of British Columbia and finally joined Oppeheimer at Berkeley.

He and Oppie were not convinced by Landau's argument that some stabilizing effect prevents a great mass of neutrons, as in the core of a large star, from continuing to collapse indefinitely. They noted that Landau, in his formulations, had used Newton's gravitational theory instead of the one newly

[85]

developed by Einstein. From their calculations it appeared that resistance to collapse would be overcome if the mass of the neutrons exceeded seven tenths that of the sun and if, being "cold," they lacked thermal motion. Under such circumstances they would go on collapsing. The most potent factor resisting collapse within a great mass of neutrons seemed to be a "snobbishness" on the part of the neutrons, causing them to fight being squeezed together in a manner much like the "electron pressure" that keeps a white dwarf from collapsing farther. Either this formulation "fails to describe the behavior of highly condensed matter," Oppenheimer and Volkoff reported in the *Physical Review* of February 15, 1939, or "the star will continue to contract indefinitely, never reaching equilibrium."

In the latter case they noted that, following the Einstein formulation, the density would become so great and the resulting gravity so strong that time, to an outside observer, would slow down and so, too, would the observed collapse. As though anxious to avoid the "absurdity" of a black hole they said: "One would hope" to find solutions "for which the rate of contraction, and in general the time variation, become slower and slower, so that these solutions might be regarded, not as equilibrium solutions, but as quasi-static."

In other words, they hoped to show that, in such a situation, time slows so much that for everyone beyond the vicinity of the star the collapse has virtually stopped.

Six months after submitting this paper and only two years before he began to become involved in the atomic-bomb project, Oppenheimer, with another graduate student, Hartland S. Snyder, took the final step and described how such a collapse would produce a black hole. Snyder, unlike most of Oppie's students, had not come from a relatively elite background. Rather, he had grown up in Utah in what was apparently an environment of rough ways and manual labor. Oppie, himself the epitome of the intellectual aristocrat, referred to him as "the rough diamond."

Snyder, however, was a mathematical wizard, and together they tackled the factors resisting collapse. The only ways they could see that a massive star could avoid such a fate would be:

1. to spin so fast its collapse would be halted by centrifugal force;
2. to blast off (or spin off) enough outer material to bring its mass below the critical level for unchecked collapse;
3. to retain sufficient residual energy, such as radiation or motion of particles in the star, to resist the pressure.

Such factors could spare some stars that might otherwise collapse, but it was doubtful that the truely massive stars could escape in any of these ways.

As the star collapsed to greater and greater density, the two men wrote, its light and other emitted energy would be increasingly reduced by three factors: One was the so-called Doppler effect, which lowers the pitch of a horn moving away from the observer. In this case inward motion of the collapsing light source would lengthen the wavelengths of light, shifting them toward the red and therefore reducing their energy. Another factor would be reddening of the light waves by the increasingly strong gravity of the condensing object. The third factor would be the bending of light waves seeking to escape the object.

As demonstrated in the 1919 eclipse, gravity can bend light waves, and the stronger it is, the more they are bent. If someone were riding down on the surface of such a collapsing object, Oppenheimer and Snyder said, that person would find it increasingly difficult to send out light signals or any other form of message. At first this would apply only to signals aimed by this bold adventurer a little above the horizon. They could not escape because gravity would bend them down to remain entrapped. As the force of gravity increased, he would have to aim signals ever higher to keep the bending effect from trapping them.

To illustrate this, imagine that one is able to control the strength of gravity by turning a knob and that someone else is launching a series of rockets all of which have identical thrust. With weak gravity the rockets can be launched clear of the earth by aiming them relatively low above the horizon.

The spiral galaxy NGC 5364 in the constellation Canes Venatici, photographed through the four-meter telescope of the Kitt Peak National Observatory.

But if the force of gravity is increased the rocket will fall back to earth unless it is aimed higher. A further increase in gravity requires an even higher angle until finally it can only escape by being fired straight up toward the zenith.

Similarly, Oppenheimer and Snyder said, the ill-fated observer riding to oblivion on a collapsing star could only get a final message out by beaming it straight up. Then gravity would cut off even that escape route. Furthermore, the wavelengths would be infinitely shifted beyond the red end of the spectrum.

"The star thus tends to close itself off from any communication with a distant observer; only its gravitational field persists," they wrote. To an outsider this vanishing process would appear to take an infinite time, they added, but for someone riding with the collapse of a typical star the isolation would become total in about one day, not taking into account such slowing factors as the residual internal pressure of the star, its radiation, and its rotation. "Of course," they said, "actual stars would collapse more slowly. . . ."

An added effect, as noted later, is that once radiation can no longer escape the collapsing star, it and the energy of the enormously increasing pressure begin, through the energy-mass relationship of special relativity, to act like mass, increasing the gravity even more rapidly. The pressure thus acts to augment gravity in the final stages, accelerating the collapse instead of slowing it.

Oppenheimer and his colleagues did not refer to the end product of this process as a "black hole." They made no suggestion that such objects ever could—or would—be detected. As with the Schwarzschild radius the concept was an exercise in abstraction—perhaps little more than a mathematical game. It was for others, three decades later, to suggest from startling new discoveries that gravitational collapse may occur on a scale beyond the wildest dreams of earlier theorists.

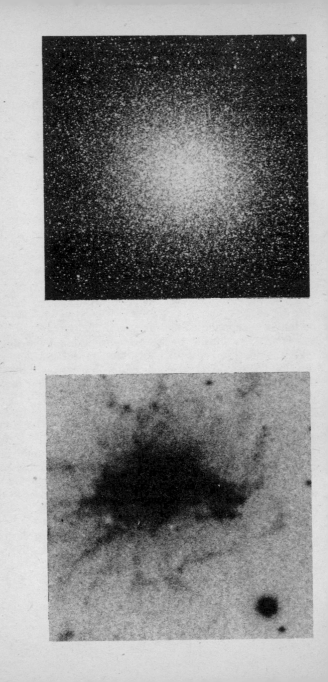

6

Cosmic Distances and Cosmic Explosions

That events of an extraordinary nature, difficult to explain in terms of conventional physics, are occurring far out in space began to become apparent in the 1950s. It was found that some objects in distant parts of the universe appear to be exploding or radiating energy in a manner millions of times more cataclysmic than a supernova.

Such discoveries became possible through the use of giant new observing instruments. In California a telescope far larger than any previously built, with a mirror 200 inches (5 meters) in diameter yet microscopically precise in its configuration, had come into operation atop Mount Palomar in 1948. Its star-tracking element weighs more than 500 tons, yet it is so perfectly balanced and lubricated that it can be moved by hand. The motor that rotates it counter to the earth's rotation, keeping its aim focused steadily on one target area among the stars, produces only one twelfth of one horsepower.

Similarly, radio telescopes of ever larger size were being built. At Jodrell Bank in England one with a parabolic, or dish-shaped, antenna 250 feet (76 meters) in diameter, fully steerable yet with a surface area almost equal to that of a football field, began scanning the heavens in 1957.

The new science of radio astronomy had been born after Karl G. Jansky of Bell Telephone Laboratories, trying in the 1930s to track down sources of static that interfered with radio communications, had found that the heart of the Milky Way Galaxy, the spiral system of star clouds in which we reside, was generating strong radio noise. We cannot see the core of the galaxy because of intervening dust, but we can guess what it is like by looking at the brilliant cores of other spiral galaxies. At radio wavelengths its "rumbling" passes through the dust without difficulty. The source of these emissions is known as Sagittarius A because it lies in the constellation Sagittarius, the Archer (The "A" indicating that it is the most powerful "radio source" in that constellation). What is going on in the core of the galaxy to produce these emissions is unknown (although some theorists now look to black holes as an explanation).

That stars also produce radio noise became evident in World War II when the operators of British antiaircraft radars along the Channel coast reported that their radar echoes were being overwhelmed by what they assumed to be German jamming equipment. The "jamming" occurred soon after sunrise—a time of special alert for air raids—but also at the time when the east-facing radar antennas were turned toward the sun. There happened to be a very active sunspot region on the sun and J. Stanley Hey, who was to become a prominent radio astronomer, realized that that was the source.

While the emissions were strong enough to jam radar sets, they were not sufficiently powerful to sustain much strength beyond the solar system, and attempts to detect radio emissions from stars other than the sun were disappointing. Yet systematic surveys of the sky, initiated in 1938 by Grote Reber with a home-made antenna in his back yard near Chicago, showed hundreds of points from which radio waves were coming. Several of these sources were remarkably strong. The most powerful of all was Cassiopeia A, whose emissions, it was found, come from a shell of gas expanding explosively from the site of a former supernova. Its outward velocity in all directions seems to be uniformly about 7,400 kilometers per second. From calculations in which this motion is reversed, like running a moving picture backward, it appears that the supernova should have been seen on earth in about the year 1700, but there is no record of it, probably because dust clouds intervened. Cassiopeia A, lying within the Milky Way Galaxy at a distance of 11,000 light-years, is "loud" because, among such supernova remnants, it is young and full of energy.

It was identification of the second strongest source, Cyg-

Edwin P. Hubble, whose observations confirmed the uniform expansion of the universe, in the observer's cage of the 200-inch (508-centimeter) telescope on Mount Palomar. A portion of the telescope's structure and dome can be seen reflected in the mirror, lower left. This giant instrument dominated observational astronomy from 1948 until a new generation of instruments, including the Soviet six-meter reflector, came into operation in the late 1970s. (*Hale Observatories Historical Photograph*)

nus A, that created a sensation in the sister worlds of physics and astronomy. The clues accumulated in small steps. Immediately after World War II Hey, using a modified antiaircraft radar, found that strong emissions, variable on a time scale of minutes, were coming from the vicinity of Cygnus, the Swan (so called because, with imagination, one can see its stars as a flying swan with extended wings and trailing legs).

John G. Bolton, another pioneer in radio astronomy, and his colleagues in Australia then used the sea as an observing aid and showed that the emissions were coming from a relatively small area, no more than eight minutes of arc in width. Each of the 360 degrees of a circle is divided into sixty minutes of arc, and each minute consists of sixty seconds of arc. The angular width of the moon is roughly thirty-two minutes—four times the upper limit set by the Australians for the Cygnus A source.

The technique they used, known as interferometry, was becoming an essential tool in identifying both the positions and apparent sizes of radio sources. Radio waves, while part of the same electromagnetic spectrum that includes light, are thousands and millions of times longer. It therefore takes an extremely large reflector to obtain a clear image of the shape and position of a radio source. The smallness (in angular width) of what can be seen is determined by the ratio of the wavelength observed to the size of the reflector. To see the "radio sky" (at, say, a wavelength of fifteen centimeters) as clearly as one sees the star-studded sky at visible wavelengths with an unaided eye would require a dish about six hundred meters in diameter.

To some extent interferometry provides a way around this handicap. The method, in which signals are recorded simultaneously by two or more antennas a distance apart, is a direct descendant of the famous experiment conducted by Thomas Young, the English physicist and physician, in 1801, proving the existence of light "waves." Ever since Newton's day it had been assumed by many scientists including Laplace that light consists of particles or "corpuscles," yet there were also those who believed in light "waves." Since light passes readily through an apparent vacuum, adherents of the wave theory supposed there must be some undetected but universal medium—an "ether"—through which such waves flow. It

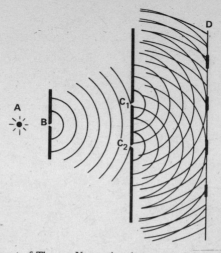

The experiment of Thomas Young that demonstrated the wavelike nature of light and the interference phenomenon at the basis of modern interferometry. Light from a point-like source (A) passes through a narrow aperture (B) forming an ordered series of wave fronts, shown as concentric curved lines. These pass through two more apertures (C_1 and C_2), producing separate wave fronts that converge on a screen (D). Where the converging waves are "in phase"—that is, where "crest" meets "crest"—they reinforce one another, forming a bright line. Where they are out of phase ("crest" meets "trough," so to speak), they cancel one another or "interfere" and no light is observed: Light added to light produces darkness.

was eventually shown that, in a sense, both schools were right. Light consists of particles (now called photons) that move in a wavelike manner.

Young demonstrated its wave nature by showing an effect seen on water when waves are arriving from two directions. In some places their crests coincide, augmenting one another. Between those spots the crest of a wave coming from one direction meets the trough of a wave from the other source. They cancel one another or, as the physicists say, they "interfere."

In Young's experiment he allowed light from the sun to pass through a pinhole, forming a beam that fell onto a panel with two parallel, closely spaced slits. The beam, split in two by passage through the slits, then fell on a white screen. Be-

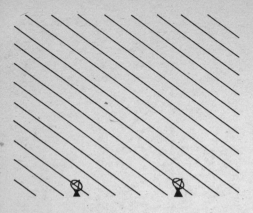

Figure 1-A

Figure 1-B

Figure 1-C

cause the original pinhole beam was small, its light waves were to a large extent "coherent"—that is, they were marching in step, like a column of soldiers. Once the beam had been split and then met on the screen, the converging waves at some points reinforced one another, producing a bright band, and at alternate positions canceled one another, producing a dark band. Thus Young demonstrated to his astonished contemporaries that, by adding light to light, it is possible to create darkness. His experiment generated a characteristic "interference pattern" of alternate dark and bright bands.

A similar effect can be achieved with an astronomical radio source. Instead of slits the waves pass through two antennas, set a certain distance apart, and then (by electrical, radio, or other means) are brought together in a common recording device. Since the earth rotates, the source of the waves moves across the sky from east to west. The antennas typically are placed on an east–west line and follow the source in its westward march. Sometimes, as shown in Figure 1-A, they both receive wave "crests" (that is, the waves are in phase). The waves reinforce one another in the receiver. But a moment later, when the source has moved, as shown in Figure 1-B, the angle of arrival has changed and a "trough" reaches one antenna as a "crest" reaches the other. Being out of phase they "interfere" and no signal is recorded in the receiver. In this way a succession of reinforcements and interferences occurs until, when the source crosses the meridian (due north or south of the site), as shown in Figure 1-C, its wave crests and troughs hit both antennas simultaneously, producing a maximum signal. Thereafter the signal becomes progressively weaker so that the record of a full passage resembles Figure 2. (The wave fronts shown in Figure 1 are essentially straight lines, since the source is extremely distant.)

While the highest, central spike in Figure 2 marks when the source crossed the meridian, helping to indicate its position in the sky, the width of each spike limits the accuracy of the observation. There are two ways to obtain more (and hence narrower) spikes. One is to widen the separation of the antennas. In Figure 1-A, which is schematic, the antennas are recording waves only four wavelengths apart. In practice the separation may run to many millions of wavelengths (antennas on opposite sides of the earth have been used). The other

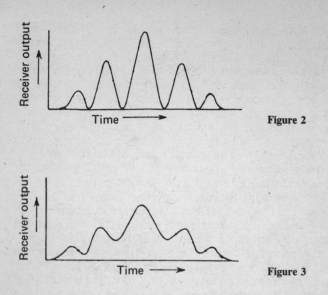

Figure 2

Figure 3

way to crowd more interference cycles into the observation is to observe at shorter wavelengths. This makes for narrower spikes, but the choice of wavelengths is limited by such factors as transparency of the atmosphere.

The same principles apply to determinations of angular width. In Figures 1 and 2 the source appears to be a point, rather than an object with identifiable width. If, however, the receiving system can discriminate between waves coming from opposite sides of the object under observation, the interference pattern will be weakened—that is, the pattern generated by waves from one side of the source will not coincide with the pattern produced by waves from the other side. Interference, therefore, will never be complete, and the signal strength, between spikes, will never drop to zero, as shown in Figure 3.

By moving antennas apart until this effect becomes apparent (or by shifting to shorter wavelengths) it is possible to determine the angular widths of objects no larger than the observed size of a star. These techniques of position and size determination became key tools in the revolutionary discoveries that were to follow, leading, for example, to construction of the so-called Very Large Array, a radio telescope on the Plains of San Augustin in New Mexico. Its twenty-seven fully steerable antennas, each twenty-five meters in diameter, are mounted on a thirteen-mile system of double railroad tracks laid out in a giant Y so the antennas can be moved to form a wide range of spacings.

The Australians, in their early observations of Cygnus, ingeniously used the ocean as a mirror—that is, as a second antenna. Their antenna, on a cliff edge, received radio waves from Cygnus both directly and reflected by the sea. At first rapid variations in the observed signal were thought to be originating at the source, implying that it was small and near, but the fluctuations did not coincide at widely separated observing sites. It became apparent that this was scintillation—an effect comparable to the twinkling of a star (which is produced by atmospheric turbulence). The brighter planets are large enough in angular width so that they do not twinkle, but stars do so, and the implication was that Cygnus A could not be very wide and was probably distant.

In England F. Graham Smith then used Cambridge University's new interferometer to scan Cygnus A and obtained the most precise radio source position up to that time. As customary for celestial objects, the position was given in right ascension (a counterpart of longitude on the earth's surface) and declination (degrees north or south of a line circling the heavenly sphere directly above the earth's equator and thus a counterpart of latitude). In right ascension Smith believed his position accurate to within one second of arc and in declination to within one minute (sixty seconds). By contrast the planet Mars, when nearest the earth, covers an area of sky twenty-one seconds of arc wide, so Smith's determination was extremely precise. He air-mailed the co-ordinates to Walter Baade in Pasadena, who had access to the one telescope in the world that, Smith thought, might be able to "see something" there.

Baade, it will be recalled, was the German-born astronomer at the Mount Wilson-Palomar observatories who, with Zwicky, proposed that supernovas leave "neutron stars" as their residue. As described some years later by colleagues, Baade was completely engrossed in his research: "Gesticulating, incessantly smoking, with carefully parted thin white hair, white somewhat bushy eyebrows, protruding hawk nose, Baade saw the mysteries of the universe as the greatest of all detective stories in which he was one of the principal sleuths.''

He received Graham Smith's letter with the Cygnus co-ordinates near the end of August 1951. "I really became interested," he later reported. "Up to then I had refused to be drawn into attempts to identify the Cygnus source. The positions had not been accurate enough. But I knew that with the Cambridge data something could be done."

Because the five-meter reflector had no rival elsewhere in the world, every second of every night was scheduled long in advance, and for each astronomer each night was an agonizing gamble on the weather. On the night of September 4, 1951, Baade, riding in the observer's cage suspended in mid-air at the focus of the giant instrument, first carried out his previously committed schedule. He took photographs of the Andromeda Nebula—the nearest spiral galaxy like our own—and of various bright nebulae within the Milky Way itself. Then, shortly before midnight, he found time to aim the telescope at the position indicated in Smith's letter. He took two photographs, one in blue light and one in yellow. The next afternoon, with the sun high (when observing astronomers often sleep), he developed the photographs.

"I knew something was unusual the moment I examined the negatives," he said later. "There were galaxies all over the plate, more than two hundred of them, and the brightest was at the center. It showed signs of tidal distortion, gravitational pull between the two nuclei. I had never seen anything like it before. It was so much on my mind that while I was driving home for supper, I had to stop the car and think."

A year earlier Baade and Lyman Spitzer at Princeton had proposed that collisions must sometimes occur in dense clusters of galaxies—"traffic accidents" on a monumental scale. In this way they sought to explain why some highly flattened galaxies lack spiral arms, as though swept clear of dust and gas by collision. If galaxies passed through one another, Baade and Spitzer reasoned, their stars, being thinly dispersed, would be unlikely to collide, but the dust and gas would do so. In Cygnus A Baade thought he saw such a cata-trophic encounter—an event sufficiently violent to generate the observed radio emissions. Other astronomers, notably his Alsatian born-colleague, Rudolph Minkowski, were not convinced. The idea that emissions as strong as those from Cygnus A and other such sources could be coming from far out among distant galaxies, millions of light-years away, instead of from within our own galaxy, seemed ludicrous.

When Minkowski gave a seminar talk on radio sources shortly after the Cygnus A observation, he ran through all the other theories and then presented that of Baade, according to the latter's account, "as if he were lifting a hideous bug with a pair of pincers." Minkowski's comment, as recalled by Baade, was something like: "We all know this situation: people make a theory, and then, astonishingly, they find evidence for it. Baade and Spitzer invented the collision theory; and now Baade finds the evidence for it in Cygnus A."

"I was angry," Baade told his friends later, "and I said to him I bet a thousand dollars that Cygnus A is a collision." Minkowski replied that he had just bought a house and could not afford that. "Then I sugested a case of whiskey," said Baade, "but he would not agree to that either. We finally settled for a bottle. . . ."

Several months later Minkowski walked into Baade's office and asked "Which brand?" He had spectrographic evidence that seemed to indicate a collision. Baade replied: "I would like a bottle of Hudson Bay's Best Procurable," which, he explained, "is the strong stuff the fur hunters drink in Labrador." What he got was a hip flask, not a quart as he expected and, according to Baade's perhaps one-sided account, Minkowski, on his next visit to Baade's home, finished off the little bottle, which Baade had been keeping as a trophy.

It has been argued that Minkowski had a right to "take back" his whiskey because it is no longer believed Cygnus A is two galaxies in collision. Rather it is a single galaxy with a dust band cutting its visual image in two, much like the Centaurus A source (known to optical astronomers as NGC 5128 and illustrated on the next page). But another form of "collision"—gravitational collapse on a grand scale—was soon to be much discussed as the energy generator of many radio sources.

The galaxy NGC 5128, known to radio astronomers as Centaurus A, appears to have ejected two objects in opposite directions along its rotation axis. It was for a time suspected of being two galaxies in collision, but is now thought to be a single galaxy with a band of dust obscuring much of its equatorial region, where violent activity is evident. Photographed through the five-meter telescope on Mount Palomar. (*Hale Observatories*)

URSA MAJOR

15,000 KM/SEC

CORONA BOREALIS

22,000 KM/SEC

BOÖTES

39,000 KM/SEC

HYDRA

61,000 KM/SEC

Twin calcium lines in the spectra of four galaxies appear at progressively greater red shifts because of increasing recession velocities, indicating greater distances. In each case the spectrum is sandwiched between spectral lines from a stationary light source. The top spectrum is from a galaxy in Ursa Major (the Great Bear), receding at fifteen thousand kilometers per second. Below it is the spectrum from a galaxy in Corona Borealis, receding at twenty-two thousand kilometers a second. The third, in Boötes, is flying from us at thirty-nine thousand kilometers per second, and the fourth, in Hydra, at sixty-one thousand kilometers per second—a large fraction of the speed of light. While the distance yardstick is uncertain, it must be several billion light-years away. (*Hale Observatories*)

While it was obvious that Cygnus A was very far away, to find out how distant the yardstick used by astronomers—the so-called red shift—had to be applied. When light from a star, galaxy, or other object is passed through a spectroscope, it is broken up into its component wavelengths, as is sunlight by a prism or by raindrops when a rainbow is formed. The resulting spectrum, if sufficiently magnified, may show thousands of bright and dark lines. The bright lines represent wavelengths of light emitted by various atoms in the light source as they jump between energy states. The dark lines are those wavelengths that have been absorbed by atoms at the source—for example, by atoms in the atmosphere of the sun.

Since virtually all galaxies are moving away from us as part of the general expansion of the universe, these spectral lines are shifted toward longer wavelengths—that is, toward the red end of the spectrum. This is the familiar "Doppler effect." It is far more commonly observed than the red shift caused by very powerful gravity and, in astronomy, is used as the yardstick for distance measurements to other galaxies. The more distant a galaxy (or cluster of galaxies), the faster it is moving away from us and therefore the greater its red shift.

This relationship between distance and speed of separation occurs, for example, in an expanding cloud of gas. In such a cloud particles of dust, because of the expansion, are all receding from one another, the rate of separation between two particles being determined by the distance between them. Suppose, for example, that a smoker blows his smoke into a balloon, partially inflating it, then heats the air to further expand the balloon. The particles of smoke in that expanding volume of air move from one another uniformly. A microbial astronomer on one particle deep within the cloud would see the same pattern of expansion in every direction. All other specks would be receding. Furthermore, if the expansion were uniform throughout the cloud, specks twice as far away would be receding at twice the velocity.

This is the nature of the universe that we see. It is the chief reason for believing it had an explosive beginning—a Big Bang—ten billion to twenty billion years ago. That the yardstick is valid is strongly indicated by the systematic relationship between the brightness of distant galaxies and their red shift. The dimmer they are, the greater their red shift in exactly the manner to be expected if the red shifts are measures of distance. In the case of Cygnus A its distance estimated in this way proved to be five hundred million light-years—so far its light has taken five hundred million years to reach us. (The actual distance may be greater, depending on the nature of space curvature and the rate of expansion, which remain uncertain.) Despite this enormous distance Cygnus A is the second most powerful radio source in the sky. Astronomers were dumfounded. How could anything so far away "shine" so brightly in radio wavelengths? Only an explosive process of some sort seemed a plausible explanation, and even that appeared to lie beyond the range of any known form of energy.

Another remarkable radio source was 3C 295. The designation means that it is No. 295 in the third (or 3C) *Cambridge Catalogue of Radio Sources,* compiled by radio astronomers at Cambridge University in England under Sir Martin Ryle (who later won a Nobel Prize for his innovations in observing techniques). The emissions of 3C 295 were shown by interferometry to be coming from a very tiny point in the constellation Boötes, the Herdsman—a source region one hundred times narrower than that of Cygnus A! Its position was located within about twenty seconds of arc—a patch of sky no larger than the apparent size of the brightest planets and so small that even through the most powerful telescope it was unlikely to contain more than a very few visible objects.

When Minkowski heard that the location was that closely established he turned to the newly completed Palomar Sky Atlas. The latter is not a book but a collection of more than one thousand high-quality photographs of all parts of the sky visible through a specially designed telescope on Mount Palomar. The eight-year project had been completed in 1957 with financial help from the National Geographic Society. On nights when seeing was excellent each field of view had been recorded twice—once in blue light and once in red—to provide clues from opposite ends of the visible spectrum to the nature of each object.

Minkowski could find nothing in the Sky Atlas photographs at the given location of 3C 295, so, like his colleague Baade, he used some of his assigned observing time on the five-meter telescope to take a long exposure of the area. From this it appeared that the emissions were coming from a faint and obviously very distant cluster of galaxies. To collect sufficient light for a spectrum, Minkowski then aimed the great telescope at the target for four and a half hours. The result showed the largest red shift observed up until that time, indicating recession at 36 per cent of the speed of light. This meant (pending precise definition of the red-shift yardstick) that 3C 295 is 10 times farther away than Cygnus A and that its light has been traveling toward us for about five billion years. We are therefore seeing it as it was when the universe was far younger than now.

The horizons of human knowledge were expanding at a breathtaking rate and every new discovery seemed to raise new questions. A number of the radio sources were soon identified as "radio galaxies" like 3C 295. And evidence was accumulating that many galaxies are undergoing catastrophic eruptions—or have done so. As early as 1943 C. K. Seyfert of Vanderbilt University had noticed that some sort of highly energetic process was at work in the cores of certain galaxies (subsequently designated Seyfert galaxies), and now the evidence for explosive processes was becoming far more awesome. That explosions on a lesser scale may take place in our own Milky Way Galaxy was hinted at by the discovery of Dutch radio astronomers that hydrogen clouds seem to have been ejected upward from its core. The escape route from flattened spiral galaxies like ours seems to be "up" and "down" because dust, gas, and stars in the flattened plane of the galaxy impede escape in any other direction. A careful examination of the Palomar Sky Atlas showed fifty to a hundred galaxies that seem to have been blown apart.

A number of "wild" explanations were offered, since no conventional one seemed adequate. It was suggested that matter and antimatter were combining and being fully converted into energy; or that stars in the cores of some galaxies are so crowded that when one erupts in a supernova, a chain reaction of continuous supernovas is started. Viktor A. Ambartsumian, head of the Byurakan Observatory in Soviet Armenia, having discounted the colliding-galaxy explanation for Cygnus A, proposed that some special physical process must occur in the cores of galaxies, a view shared by several other astrophysicists.

Not only was the energy source a puzzle, but so was the manner in which light from some explosion features was being generated—notably in the Crab Nebula. As pointed out earlier, that rapidly expanding gas cloud is the residue of a supernova explosion observed in A.D. 1054. It is the third most intense radio source in the sky. While light from filaments in its rapidly expanding bubble of gas can be broken down into spectral lines, much of the light comes from an amorphous cloud that, unlike the light from most celestial sources, shows no such lines. Theorists were doing speculative handsprings trying to explain why this might be.

Baade and Minkowski had suggested that highly ionized gas was being made to glow by a very hot star, left from the supernova, but there were problems with this explanation, and in 1953 Iosif S. Shklovsky and Vitali Lazarevich Ginzburg in the Soviet Union proposed that the emissions were a form of radiation discovered accidentally only six years earlier.

Engineers at the General Electric Research Laboratory in Schenectady, New York, had learned that Edwin M. McMillan at the Radiation Laboratory of the University of California in Berkeley was planning to build a new and far more powerful kind of electron accelerator (it came to be known as a synchrotron). They decided to have a go at it themselves and while testing the machine heard ominous noises inside it, as though of sparking. Herbert C. Pollack was there and has described the scene. It was decided to rig a mirror so that they could look around the protective concrete wall without exposure to the X rays they assumed were being generated by electron collisions. With this, on April 24, 1946, Floyd Haber, a technician, looked into the racetrack around which the electrons were flying, held in their circular path by strong magnetic fields. "Turn off the machine!" he cried, for he saw a brilliant blue light and assumed it was electrical arcing.

It was not. It was the first observation of what has come to be known as synchrotron radiation—unlike all other forms of light in that it has no spectral lines. It does not follow the rules of quantum behavior, its wavelengths being controlled only by the energy of the circulating electrons that is shed in the form of light waves (or other waves) as the magnetic field pulls them around corners. The possibility of such radiation

The Great Spiral Nebula in Andromeda, nearest spiral galaxy like our own Milky Way system.

The spiral galaxy NGC 5194 in the constellation Canes Venatici (the Hunting Dogs) with a satellite galaxy, possibly ejected from the larger one. Photographed by the five-meter telescope on Mount Palomar. Such galaxies contain a brilliant core whose energy source is hotly debated. (*Hale Observatories*)

The spiral galaxy NGC 4565 in the constellation Coma Berenices, seen edge-on with typical bulge in the center. (*Kitt Peak National Observatory*)

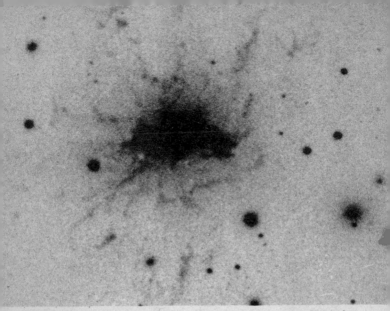

NCG 1275, an "exploding galaxy" photographed by C. Roger Lynds from Kitt Peak at the red wavelength known as hydrogen alpha and printed in negative. (*Kitt Peak National Observatory*)

had been suggested a half century earlier and it had been looked for in weaker accelerators since it would account for the energy losses sustained by electrons circulating in a strong magnetic field. But physicists had miscalculated the relationship between the energy of the electrons and the resulting radiation wavelengths.

Shklovsky and Ginzburg realized that if electrons in the Crab Nebula were spiralling around powerful magnetic fields at close to the speed of light, that could account for the nebula's strange glow. The idea could readily be tested because the light from synchrotron radiation is strongly polarized— that is, the orientation of light-wave oscillations tends to be uniform. Soviet observations soon identified such polarization. While the Palomar astronomers apparently did not put much stock in the finding, Baade finally decided to observe the Crab Nebula with the five-meter instrument and, sure enough, its light was strongly polarized. Baade then found that, as Shklovsky had predicted, the same kind of light was coming from a bright blue jet of material blasted out of galaxy M 87 in the constellation Virgo, the Virgin, suggesting elements in common between the explosion of a star—a supernova—and explosions in the core of an entire galaxy.

By the early 1960s such radiation was being observed from many features of the ''exploding'' galaxies, but this did not explain the energy source responsible for those eruptions. Among those who sought an answer was a somewhat unorthodox quartet of astronomers and astrophysicists who had begun working together at Cambridge University in the mid-1950s. One was William A. Fowler, on sabbatical leave from the California Institute of Technology and an authority on how the heavier elements came to be formed as the universe evolved. The others were all Britons: Geoffrey Burbidge; his wife, Margaret; and Fred Hoyle. At a scientific meeting Margaret Burbidge, despite her soft-spoken manner and almost angelic appearance, was adept at tearing to shreds the more inept proposals of other theorists. She later became the first woman to head Britain's Royal Greenwich Observatory, from which all time zones and longitudes are reckoned. Her husband, Geoffrey, became director of the Kitt Peak National Observatory in Arizona. Fred Hoyle was not only an originator of the ''steady state'' cosmology—the idea that the universe has always been as it is today—but also a popular science-fiction writer (*The Black Cloud* and other books).

The group began looking at the possibility that gravitational collapse on a much larger scale than that of a supernova might be the energy source of the violent events being observed at great distances. Even though gravity is by far the weakest of the forces in nature, when it draws together a large amount of material, far more energy can be released than by other means, including the various nuclear processes. Such a collapsing together, in a sense, retrieves some of the energy of the original Big Bang. The energy that blew the universe apart is, in a local region, recaptured. Having been ''potential'' energy it becomes ''kinetic'' energy on an immense scale.

Early in 1963 Hoyle and Fowler published two papers discussing the possibility that ''supersupernovas'' involving stars perhaps a hundred million times more massive than the sun might be the answer. If a supermassive star collapsed,

The author and Margaret Burbidge.
(*University of Miami*)

Sir Fred Hoyle
(*Floyd Clark, Cal Tech*)

they reasoned, as its spin energy (angular momentum) became concentrated into a small volume it would begin spinning so fast that it would throw off huge blobs. They might be as massive as ten million suns and moving almost at the speed of light. This came close to what was being observed in the jets shooting out in opposite directions from some galaxies. In fact, when those galaxies were mapped at radio wave lengths, it was often found that most of the radio energy is coming, not from the galaxy itself, but from two areas far out on either side of it. That the explosion responsible for the typical jet was not just a "flash in the pan" was indicated by the jet's great length, indicating a process going on for thousands or even millions of years.

Hoyle and Fowler's 1963 discussion of gravitational collapse on a mammoth scale as a possible source of energy for such events was a prelude to what would follow later that year with the culmination of discoveries that were probably the most extraordinary since Galileo in 1609 first turned a telescope on the skies.

7
Quasars

The first hints of the discoveries that made 1963 memorable came in the preceding three years, when it was found that several strong radio sources could not be associated with galaxies. Among them was one listed in the Cambridge catalogue as 3C 48.

The largest radio telescope then operating, at Jodrell Bank, had been paired with a movable array to form a tandem observatory—an interferometer like those described in the preceding chapter. Observations at four antenna separations, ranging from 40 to 110 kilometers, showed the width of 3C 48 to be less than four seconds of arc—comparable to that of Mars at its greatest distance. At the California Institute of Technology Thomas A. Matthews then used the institute's twin ninety-foot dishes to obtain as good a position as possible. The dishes are in Owens Valley, and relatively remote from automobile ignition systems and other sources of radio interference. Finally, Allan R. Sandage of the Mount Wilson-Palomar observatories (operated jointly by Cal Tech and the Carnegie Institution of Washington) took a ninety-minute exposure of the area with the five-meter reflector. He found what seemed a very faint bluish star with a wispy feature nearby.

To see what kind of star it was, the spectrum of its light was recorded, and Sandage's colleague, Jesse L. Greenstein, found it very peculiar. No evidence of hydrogen, the chief constituent of normal stars, could be identified. Some spectral lines appeared near the characteristic wavelengths of helium, ionized calcium, and possibly highly ionized oxygen (in an ionized atom one or more electrons have been "boiled off"). The brightness of the object varied in a manner that seemed to show it was a nearby star rather than a distant galaxy. The idea that a galaxy formed of a hundred billion stars spread over a vast volume of space could vary in brightness in a systematic manner was preposterous. Variability of such a galaxy, or anything else that large, would be equivalent to switching on and off simultaneously the light of a hundred billion suns. This, one group of theorists said, seemed "utterly ludicrous," and it was assumed that 3C 48 was within the Milky Way. In that case it would not be in sufficiently rapid motion, relative to the earth, to alter substantially the wavelengths of its lights.

When these findings were reported in a last-minute paper offered at the New York meeting of the American Astronomical Society in December 1960, it was assumed a new kind of star had been found—the first observed to be emitting enough radio energy to be "heard" over large distances.

The five-meter reflector was also aimed at two other radio sources with no obvious optical counterparts—3C 196 and 3C 286. In neither case could even the faintest galaxy be seen. Yet near each indicated position was an odd-looking star rich in blue, violet, and ultraviolet wavelengths. Most of the light from these objects lacked any spectral lines at all, and those that could be detected were broad, fuzzy lines of emission.

The break came when Cyril Hazard and his colleagues in Australia used the moon and their new 210-foot dish antenna to pinpoint another strong radio source—3C 273. Between April 15 and October 26 of that year (1962), they calculated, part of the moon would pass three times across the region occupied by 3C 273, cutting off its emissions for a certain length of time.

Since the configuration of the moon's disk, as seen from earth, is very precisely known, as is its path across the sky, the moment when the signals were cut off and when they were again received would provide very accurate information on where their source lay among the stars. The cutoff time (occultation) would be longest if the source was obscured by passage of the full width of the moon. If, instead, the time was shorter, its duration would indicate how much of the moon's upper or lower half had crossed the source—and therefore would pinpoint the source's location. While on the first pass it might not be possible to tell whether it was the upper or lower half of the moon that passed across the source, this uncertainty should vanish after three such crossings along slightly different paths. The experiment would also show if the source was an extended, fuzzy object or closer to being a pinpoint in the sky. In the former case it would fade gradually as the moon covered more and more of it. In the latter case it would vanish more abruptly, its signal fluctuating due to diffraction (bending of the light waves around the edge of the moon).

The Australians found that the source was actually in two parts. Component A was the stronger of the two and showed a radio-wave spectrum typical of many other radio sources. The spectrum of Component B, they reported, was "most unusual, no measurement of a comparable spectrum yet having been published." With the positions of each known with greater precision than any other radio source, the Palomar astronomers aimed their giant scope and found a starlike object precisely at the location of Component B—the unusual one—and "a faint wisp or jet" associated with Component A. The jet was separated from the starlike object but was aimed directly away from it, as though it had been blasted out of the "star." Component A was at the outer tip of the jet.

Optically the starlike radio source was the brightest yet observed—bright enough to provide a better spectrum than had been obtained for 3C 48. Maarten Schmidt, a Dutch-born astronomer noted for his painstaking analyses of spectra, aimed that indispensable instrument, the five-meter telescope, at the star and obtained a spectrum that, while fuzzy, showed hints of six emission lines—smeared out into broad bands—superimposed on uniform blueness.

"The only explanation found for the spectrum," Schmidt reported after testing many possibilities, "involves a considerable red shift." He identified four of the lines as those emitted by hydrogen at a harmonic succession of wavelengths (known as the Balmer series, for the Swiss who recognized their mathematical relationship). If it was the nucleus of a galaxy whose red shift was caused by motion away from the earth, its rate would be close to fifty thousand kilometers per second, or one sixth the speed of light. This, Schmidt said, implied a distance of about 1.6 billion light-years (it could be twice that far, depending on the still uncertain nature of the universe's expansion) and a source of light one hundred times brighter than any of the radio galaxies found so far.

He assessed the possibility that, instead, the source was an object so dense that the red shift was caused by its gravity. The gas surrounding such an object, however, would not display a single, well-defined gravitational red shift. The inner region would be in a far stronger gravity field than the outer part, the result being a smeared-out succession of red shifts. It would thus be "extremely difficult, if not impossible," he said, "to account for the red shift in this way."

The Palomar group then went back and took another look at the spectrum of 3C 48, which they had identified as a new kind of radio star in our galaxy. They found that it could better be explained if one of the emission lines was produced by ionized magnesium at a wavelength so short that it normally occurred in the invisible (ultraviolet) part of the spectrum. Rocket observations of the sun from above the atmosphere have shown this line to be the brightest component of solar ultraviolet, but here it was apparently red-shifted to such an extent that it appeared as visible light.

This implied a receding velocity of 110,000 kilometers per second or more than twice that of 3C 273. It indicated a distance of more than three (and perhaps as great as seven) billion light-years and, reported Matthews and Greenstein of the Cal Tech-Palomar group, could be interpreted as "the central core of an explosion in a very abnormal galaxy. . . ."

These developments came in such quick succession that they were all reported together in the March 16, 1963, issue of *Nature:* an account of the Australian use of the moon as an observational aid, the identification by the Palomar group of a "star" at the derived location, Schmidt's determination of its extraordinary red shift, and the reassessment of 3C 48. To astronomers around the world the news was like a bugle call or, better yet, a starting gun for the race to find out what these remarkable objects were.

The quasar 3C 273 with its associated jet. The radio emissions come from the quasar itself and from an object at the outer tip of the jet. Photographed through the four-meter telescope of the Kitt Peak National Observatory.

[121]

The California astronomers referred to them as "quasi-stellar radio sources"—a term far too cumbersome for the fast-moving investigations and debates. Hong-yee Chiu at the Goddard Institute for Space Studies in New York suggested a short form—"quasar"—which to astronomers of a conservative bent seemed too jazzy. But soon it was part of the scientific vernacular, although not until 1970 did Chandrasekhar, as editor of the *Astrophysical Journal,* permit its use there. The first time he did so (in an article by Maarten Schmidt) he appended a footnote: "The *Astrophysical Journal* has up till now not recognized the term 'quasar'; and it regrets that it must now concede: Dr. Schmidt feels that, with his precise definition, the term can no longer be ignored."

The brightest of the quasars, 3C 273, showed up on numerous sky photographs, although no one had paid it much heed, assuming it was just another among millions of faint stars. Now it became the focus of intense interest. Harlan J. Smith and his colleagues at Yale (from which he then moved to the University of Texas) went through some five thousand photographs of that region of the heavens dating back to 1886. He found that the brightness of the object was highly variable. Since 1929, in particular, it seemed to have faded and grown brighter in a cycle of about thirteen years and also to have flared up on time scales of only a few days. In fact, some quasars radically change their output of energy within hours.

If 3C 273 is really "nearby"—a star within our Milky Way Galaxy—then it must participate in the grand rotation of the galaxy. The spiral of star clouds that forms our home in the universe is like a giant merry-go-round, our part of which rotates once every 250 million years. None of its stars, observed over a long period, therefore remain stationary relative to distant galaxies. They display what astronomers call "proper motion." William H. Jeffreys of Yale checked through old photographs for evidence of such proper motion by 3C 273 and found that it seemed absolutely stationary.

Before 1963 was out, nine quasars had been identified optically. One, 3C 286, seemed to be receding at 55 per cent of the speed of light, making it even more distant than the radio galaxy 3C 295 and thus by far the most distant object known. This could be taken to mean that the quasar is ten billion light-years away. At such great red shifts, however, the distance estimate depends on assumptions regarding the universe—namely the rate of its expansion and the extent to which it has slowed since the period following the Big Bang. We are seeing 3C 286 as it was billions of years ago when the expansion rate presumably was faster. How much faster is uncertain. Some astronomers assigned more of the red shift to this effect and put the distance to 3C 286 as closer to six billion light-years.

By June of 1963 excitement—and perplexity—within the scientific community had reached such a pitch that a group of specialists in relativity theory decided to call a meeting of the world's leading theorists and astronomers to exchange ideas and observations. (Those issuing the invitation were Peter G. Bergmann of Yeshiva University in New York, Ivor Robinson of the Southwest Center for Advanced Studies in Dallas, and Alfred Schild and E. L. Schucking, both of the University of Texas.)

"For more than ten years," they wrote, "the nature of the strong extragalactic radio sources has been one of the most fascinating problems of modern astronomy. For a time, it was believed that such radio sources were due to collision of galaxies. But it has emerged in the course of the last few years that this explanation is untenable in most of the cases. The spectacular nature of strong radio sources becomes clear if one considers the enormous amount of energy involved." This, they added, "has so far ruled out nearly all of the explanations and theories put forward to explain such extraordinary events."

In February of that year (1963), their invitational letter continued, Fred Hoyle and William Fowler had suggested "that energies which lead to the formation of radio sources could be supplied through the gravitational collapse of a superstar. Such an object, with a mass between one hundred thousand and one hundred million solar masses, would be located in the center of the galaxy. The gravitational collapse of this superstar could supply the necessary energy if it were to shrink down close to the Schwarzschild radius." They cited the identification of 3C 273 and other recent developments and proposed that a special conference be held in Texas late in the year.

It was to be in Dallas, under the imposing title, "An International Symposium on Gravitational Collapse and Other Topics in Relativistic Astrophysics." Four federal agencies—the Air Force, the Navy, the National Aeronautics and Space Administration, and the National Science Foundation—agreed to provide financial support. Then, on November 22, as the scientists were preparing for their trip to Dallas, John Fitzgerald Kennedy was assassinated a few blocks from where the meeting was to take place. So intense was the feeling of horror that some scientists announced they would not go to Dallas, but in the end almost all of those invited came—apart from Shklovsky and his colleague, Vitali L. Ginzburg, who apparently were not allowed by the Soviet authorities to attend.

For almost a quarter century Robert Oppenheimer's black-hole proposal had been collecting dust among back copies of the *Physical Review*. As head of the Institute for Advanced Study in Princeton he was now more administrator than theorist. But suddenly what had seemed an intellectual exercise remote from reality—his calculations on the destiny of gravitational collapse—had been thrust into the forefront by efforts to explain the quasars.

When Oppenheimer arrived to serve as chairman of the opening session he looked even more long, lean, and ascetic than in his younger years. Also on hand was Martin Schwarzschild of Princeton University, son of the man who first described the Schwarzschild radius, theoretical core of the black-hole concept.

The participants noted that nuclear energy, such as that which makes the stars shine, was a poor candidate as the quasar energy source. To produce the observed emissions of light and radio waves—assuming distance estimates were correct—one hundred million suns would have to burn up their nuclear fuel in one hundred thousand years. Most star lifetimes are reckoned in billions of years. Furthermore, Schwarzschild noted, no assemblage of matter comparable to one hundred million suns has ever been observed working "in a concerted manner."

The possibility that huge stars were involved seemed unlikely to some. Because of radiation pressure within any star more than one hundred times as massive as the sun, the star, it was argued, would pulsate so wildly that much of its material would be thrown off. It was also proposed that the elapsed time from when collapse of a superstar had proceeded far enough to produce the observed radiations until everything vanished inside the Schwarzschild radius would be very short—only about one day.

Hoyle and Fowler nevertheless restated and elaborated their idea of collapse on a vast scale. To introduce the published version of their paper they borrowed a statement made in San Francisco by Bob Fitzsimmons before his (unsuccessful) attempt in 1902 to regain the world heavyweight boxing championship from James J. Jeffries, a very big man: "The bigger they come the harder they fall."

The type of object Fowler and Hoyle had in mind was more massive than a million suns, but concentrated in a small area, astronomically speaking (about thirty-five cubic light-years). "For the moment," they wrote, "we ignore the question of how such an object might be formed—the observational evidence would seem to give strong support to the postulate of the existence of massive objects, and it is therefore reasonable to inquire into their properties without further ado. We turn a blind eye, a deaf ear, and a cold shoulder to written, oral, and implied criticism, respectively."

With the Burbidges they also proposed that total collapse is prevented by a form of antigravity that Hoyle believed was causing expansion of the universe. Hoyle argued for a "steady state" universe in which there had never been a Big Bang; instead, expansion was caused by the effects of "negative energy," or what he called the "C-field," which pushed things apart just as gravity pulls them together. This, it was proposed, would prevent the superstar from collapsing to the state known as a "singularity," where mass and energy are concentrated into an infinitesimal point, where space vanishes and time comes to an end—a concept that makes a number of physicists uneasy. Instead, some of the collapsing material, after it had shrunk inside the Schwarzschild radius, might bounce back, re-emerging to help generate the observed radiation. The bouncing effort would eventually subside and the entire superstar would vanish inside the radius of invisibility, the four theorists proposed. Such huge, invisible objects, they said, may be adrift among the clusters of galaxies, providing the gravity needed to hold them together. It was unclear why galaxies occur in clusters instead of having been spread randomly throughout the universe by their rapid motion. Only a fraction of the mass needed gravitationally to hold them together in clusters seemed to be observable.

The most articulate proponent of some form of black-hole explanation was John A. Wheeler of Princeton University, who, after participating in theoretical work that led to the first thermonuclear (hydrogen bomb) explosion in 1952, began teaching a course in general relativity. To him the most fascinating aspect of that theory was its implication that collapse can go all the way to a singularity. He was not convinced by the proposed brakes to prevent it, and it was he who invented the term "black hole." As an exercise in abstract analysis much like Karl Schwarzschild's study of point sources of gravity, Wheeler had studied objects formed entirely of energy—that is, electromagnetic radiation—and containing so much of it they would be held together by their own gravity. Since energy and mass are interchangeable, energy produces gravity just as mass does. But it takes a great deal of energy to produce even weak gravity, since the equivalent amount of mass (and therefore gravity) is equal to the energy divided by the speed of light squared.

Wheeler called his hypothetical ball of self-gravitating energy a *Kugelblitz* (the German term for ball lightning) or a "geon." In a quasar, he reasoned, once matter has fallen inside the Schwarzschild radius, it may become rather "geon-like." A factor causing the collapse to run even faster, once everything has fallen inside that radius, is the inability of energy to escape. Before the material vanishes inside the radius, energy radiation helps serve as a safety valve, but afterward nothing can depart, and as the black hole shrinks, the radiation adds to the gravity, as in a geon, causing the collapse to proceed more rapidly.

The theorists and observing astronomers at Dallas dug deep into their imaginations. They discussed collisions between normal galaxies and those formed of antimatter. They talked of twin superstars spiraling in toward one another, releasing gravitational energy as they did so. They proposed that when a system of many stars collapses, its rotation rate becomes so rapid some of the stars are thrown off, carrying

[126]

with them much of the angular momentum. This would eventually leave a tight, slow-rotating cluster of stars like the so-called globular clusters.

Objections were raised to all of the proposals. Collapse to within a Schwarzschild radius, it was said, would be halted by fast rotation or by a lack of symmetry in the collapsing material. Peter Bergmann, co-organizer of the meeting and an authority on general relativity, warned his colleagues in a final summation: ". . . let us remind ourselves that the theory of general relativity does not represent the ultimate truth, any more than does any other physical theory." As Oppenheimer put it, "almost nothing" is known of what happens in the presence of extremely strong gravitational fields.

But the idea that some new form of physics must be invented did not appeal to Philip Morrison, then at Cornell University, who did another of the summations. "Some participants," he said, "notably Hoyle and Wheeler, have been bold enough to speculate on completely new physics at the basis of these events. I remain interested but unpersuaded." He nevertheless expressed the feeling of many at the conference: "The sense of wonder and excitement that has been generated is amazing."

The problem of the quasars, far from going away, has proved stubborn. Conferences modeled after that historic one in 1963 have been held repeatedly—first once a year and then less often. They have come to be known as the Texas Conferences on Relativistic Astrophysics, although not always held in Texas (that in 1978 was in Munich). A recurring question has been whether the quasars are really as distant as assumed, for, if they are not extremely far away, their energy production is not such a puzzle.

For normal galaxies, including the strong emitters of radio energy (the radio galaxies), red shift has proved a reliable indicator of relative distance. This is confirmed by the observation that at greater red shifts they become systematically dimmer. For quasars taken as a whole, however, this is far less true. The reason appears to be that they vary radically in their intrinsic luminosity, which (at least without some sort of correction) cannot therefore serve as an indicator of relative distance. In that case the red shifts would still be a valid yardstick. It has, in fact, been observed that some quasars undergo extreme changes. William Liller and his Harvard colleagues have plotted the brightness of 3C 279, a highly variable quasar (and at times the brightest) as recorded in photographs from 1929 to 1952. It varied by a factor of more than 480, increasing in brightness almost eightfold during thirteen days in 1936. Some quasars seem to fluctuate in brightness on time scales measured in hours.

Such rapid fluctuations remain a major puzzle, setting stringent limits on the size of the energy source. An object like a star or galaxy cannot systematically vary its energy output, be it in light waves or radio waves, on a time scale shorter than the length of time required for light to traverse that object.

For example, since it would take a light signal—or any other such phenomenon—at least 4.6 seconds to traverse the diameter of the sun, the sun as a whole could not fluctuate on a time scale shorter than 4.6 seconds. Since the Milky Way Galaxy is more than a hundred thousand light-years in diameter, it should not behave, in any organized manner, on a time scale of less than many thousands of years.

The effect can be illustrated in terms of the touchdown shout heard from a nearby football stadium. The shout cannot be shorter than the time required for sound to cross the width of the stadium. That interval determines how much sooner the shout of those nearby arrives ahead of the shout from the most distant onlookers. In a more subtle sense—closer to the astronomical situation—the crowd itself would not shout at the same instant, since light moves at a finite speed. Those nearest the play would see it first. Those in the most distant seats would see it last and shout last—time differences inconsequential in a stadium, but of major importance on an astronomical scale. The "shouts" coming from quasars must originate in areas that, in some cases, are no larger than the solar system. And yet their recorded emissions are strong at distances assumed to be billions of light-years.

A puzzle cited by those doubting that quasars are extremely far away has been the evidence that some of them seem to have ejected gobs of material at speeds six or eight times the speed of light. Physicists almost universally regard such speeds as impossible, and it is therefore argued that the objects are closer, just as rapid motion of an aircraft across the field of view is plausible if it is relatively close.

The faster-than-light, or "superluminous," behavior is observed in terms of increasing angular separation between a quasar and an object presumably ejected from it. This is recorded by the interferometry method using widely separated radio antennas. Among proposed explanations is one by Martin Rees of Cambridge University where the motion would appear deceptively rapid. If the object were ejected obliquely toward the earth at just under the speed of light, it would be approaching the earth almost as fast as its own radio emissions. It would, so to speak, be hot on the tail of emissions it radiated a hundred years earlier. Thus those emitted at a certain time and those of a century before would reach the earth almost at the same time, making it appear that the increase in angular separation over a hundred-year period had occurred in a matter of weeks. Whereas radio galaxies such as NGC 5128 (Centaurus A) eject material in opposite directions, the jets radiating from quasars such as 3C 273 are single. If the jet we see is coming toward us almost at the speed of light, one in the opposite direction would be so red-shifted by its velocity that it would be invisible, particularly if quasars are extremely distant.

One explanation for the great quasar red shifts would be their explosive ejection from our own galaxy. This was proposed at the 1964 Texas meeting by N. James Terrell of the Los Alamos Scientific Laboratory. He pointed out that this would account for what had been taken as evidence of their extreme distance—their lack of any "proper motion." Usually one can tell immediately whether a light in the night sky is a star or a plane by whether or not it moves. But if the plane is flying directly away from the observer, its light seems as motionless as that of a star. This, Terrell proposed, is why quasars, all flying away from us, show no proper motion. Proponents of this argument pointed out that a number of "exploding" galaxies seem to be throwing out material at very high velocity. Geoffrey Burbidge and Fred Hoyle suggested that the quasars may have been ejected, not from our own galaxy, but from the relatively nearby one responsible for some of the most powerful radio emissions in the sky—

Centaurus A. Both from its radio emissions and optical appearance it is clearly in the throes of some great upheaval. Yet it was noted that if galaxies have ejected quasars in this manner, some of them should be flying toward us at high velocity and their light should be "blue-shifted." Yet none are seen. It seemed improbable that such an explosion would be unique to one galaxy—our own.

A discovery that may help astronomers determine the nature of quasars has been identification of objects with many—but not all—features typical of quasars. As with the latter, they had long been evident on photographic plates and had been assumed to be faint stars within our own galaxy—close neighbors, so to speak. An observation by the Canadian National Radio Observatory, however, showed one of them to be a radio source with a peculiar spectrum (its intensity increases at higher frequencies instead of becoming weaker, as in most sources). It was then found to be almost as distant as the quasars and therefore extremely brilliant.

The "star" had been known as BL Lacertae, the BL being a two-letter code designating it as the ninetieth variable star discovered in the constellation Lacerta, the Lizard. Some forty such "BL Lac" objects have now been discovered. Some flare to six hundred times their previous brilliance, becoming one hundred times more luminous than the entire Milky Way Galaxy. BL Lacertae itself varies fifteen fold in brightness, changing 400 per cent in two days and flickering a few per cent within minutes. The energy source may therefore be even more compact than that of a quasar—smaller than the solar system. How such torrents of energy can be generated within so small a region is baffling.

The BL Lacs seem embedded within faint elliptical galaxies, although only a few such galaxies have been observed by blocking out the brilliant light of the central object. Unlike quasars they show only very weak spectral lines and their light is strongly polarized in the manner produced by synchrotron radiation rather than heat. If such clues can lead to an understanding of their energy source, this could explain the quasars as well.

Among the most sensational developments in early quasar observations was a 1965 report that Shklovsky's group in Moscow had detected fluctuating signals from the radio source CTA 102 indicating that it was the beacon of a civilization seeking to draw attention to itself. (The CTA designation refers to the Cal Tech A catalogue of radio sources.) The initial report came from the Soviet news agency TASS. Shklovsky, then the leading Soviet proponent of the probable existence of other civilizations with a high technology, hastened to explain that the signals were not necessarily of intelligent origin. He added, however: "One cannot exclude, of course, the fascinating conjecture that what we are observing is an artificial signal from an extraterrestrial civilization. New, special observations will be needed, however, before this theory becomes a scientific fact."

The original observations had been made after it had been noted that two radio sources—CTA 102 and CTA 21—emitted their peak power in the frequency range best suited to communication between civilizations—one that readily penetrates an atmosphere like that of the earth and encounters little competition from other sources of radio noise. To see if either source was variable in a way that might carry a message, each was observed simultaneously with a source believed to be steady, namely 3C 48. In this way fluctuations caused, for example, by the earth's own atmosphere would be canceled out since they would influence both sources equally. Only changes in the relative power of the two sources would be recorded. The signals from CTA 21 were found to be unchanging, but those from CTA 102 seemed to vary in a one-hundred-day cycle. The beacon, it was proposed, lies within our own galaxy. The excitement, however, was short-lived. The Palomar telescope was aimed at the object and recorded a faint, violet star. Maarten Schmidt determined its red shift and showed it was a quasar and presumably very distant.

By 1967, when the International Astronomical Union brought the world's astronomers to Prague, more than one hundred quasars had been found. The meeting was exhilarating. Not only were a wide range of new discoveries presented but also the atmosphere in Prague itself was one of excitement. It was the "Prague spring," when Soviet control had been relaxed and the Czechs were celebrating their new freedom with special fervor—a mood that was to be cut short a year later when Soviet troops moved back in force.

During this period, since no generally accepted explanation for the quasars had appeared, more bizarre proposals were advanced. A clustering of quasars at certain red shifts had been observed, and it was suggested that they define a "crystal structure" in the universe, each placed at a key point in the symmetrical lattice of that supercrystal. Another idea was that quasars are remnants of the original fireball in which the universe was born.

Two years earlier a pair of researchers at Bell Telephone Laboratories in New Jersey had made an extraordinary discovery while working with the prototype antenna for the first satellite communications system. It was atop Crawford Hill almost within sight of the field near Holmdel where Karl Jansky in the 1930s had first observed radio emissions from the heavens. The antenna was a giant device of the horn-shaped variety mounted so that it could be aimed anywhere in the heavens. Since it had completed its role in preparations for the first transatlantic link via the Telstar satellite (its twins were operating at Andover, Maine, and at Pleumeur-Bodu, France), Arno A. Penzias and Robert Wilson had begun to use the instrument for radio astronomy observations.

In studying emissions from the Milky Way they sought to eliminate, or account for, all sources of noise in the system—emissions from its own electronics, radio noise from the atmosphere and from specific radio sources in the heavens. When all sources had been identified or corrected there was a small residue they could not explain. It was not atmospheric in origin, for then it would have become stronger when the antenna was aimed low and thus was "seeing" more air than when aimed straight up. The source was not in the Milky Way Galaxy, since the emissions were equally strong no matter where the antenna was aimed.

Pigeons, found to be nesting inside the antenna, were caught and released at a distant site. They were soon back, however, and a more permanent solution was necessary. Inside the antenna the birds had left what Penzias delicately described as "evidence of their visits," which was also removed. Still the unaccounted-for noise remained.

The experimenters remachined the flanges at the antenna throat and carefully checked other parts of the structure to insure smooth operation. They went over the circuitry with an attention to detail comparable to that used in preparing a

Arno Penzias, left, and Robert W. Wilson in front of the horn antenna with which they detected a universal "glow" in the microwave region of the radio spectrum believed to be the residue of the Big Bang fireball. (*Bell Labs*)

The bright spiral galaxy NGC 1566, showing its nucleus where explosive processes, not fully understood by astronomers, liberate immense amounts of energy. Photographed by the 3.9-meter Anglo-Australian telescope, Siding Spring, Australia.

spacecraft for manned flight. This, too, did not eliminate the extra noise. Perhaps, they thought, it was coming from automobiles passing on the nearby Garden State Parkway or some other local source, although the design of the antenna supposedly ruled this out. They carried a transmitter to various points in nearby fields, but there was no sign of leakage into the system.

The amount of unexplained noise being observed, expressed in the terms used in such work, was the radio energy at extremely short wavelengths (microwaves) that would be emitted by a totally black object heated to about three degrees above absolute zero (the total absence of heat at minus 273 degrees).

This may not sound like a very bright radio "glow" in the sky, but if, in fact, it filled the whole universe it would represent an enormous amount of energy. Penzias and Wilson then learned through the scientific grapevine that at nearby Princeton University the group led by Robert H. Dicke (including P. J. E. Peebles, P. G. Roll, and D. T. Wilkinson) were building an antenna designed to pick up just such emissions in the expectation that they would remain as a residue of the hypothetical fireball in which the universe was born. It was assumed that the flash of the primordial explosion would still be contained within the expanding universe, but that the wavelengths of its light, would have been greatly stretched as the expansion continued. The waves would no longer lie within the visible part of the spectrum but would appear in the microwave region (of radio wavelengths). Peebles had calculated that this microwave "glow" would be very faint— comparable to that emitted by a black body only a few degrees above absolute zero, the exact temperature being dependent on the manner in which the universe was born— whether, for example, from "nothing" or from what the Princeton group referred to as "the ashes of the previous cycle" in a universe that never had a beginning but that oscillates between expansion (its present state) and contraction.

In 1928 Ralph A. Alpher and Robert Herman, who had been working with George Gamow on his "Big Bang" theory, had predicted a residual glow equivalent to a temperature of five degrees above absolute zero. Yet no one had tried to observe it, and the prediction apparently was unknown both to the Bell Labs researchers and the Princeton Group. The discovery therefore came about totally by accident. In fact, as one of the Bell Labs scientists commented afterward, only when he saw an account of the discovery displayed on the front page of *The New York Times* did he realize the sensational nature of their observation.

Steven Weinberg, one of the leading theorists in physics, considers this "one of the most important scientific discoveries of the twentieth century" (for which Penzias and Wilson, in 1978, shared a Nobel Prize). Why, he asks in his book *The First Three Minutes,* was the prediction of Alpher and Herman forgotten instead of being followed up? The answer was in part because no receiving system sensitive enough to record the glow was in prospect in 1948. Weinberg suggests, as well, that so sensational an idea as looking for the Big Bang flash was considered, by a hidebound scientific community, unfit "for respectable theoretical and experimental effort."

In truth, however, it was not entirely forgotten. Shortly after World War II a young man named James W. Follin walked into the office of Allan Sandage at the Mount Wilson-Mount Palomar observatories headquarters in Pasadena. For several hours they discussed cosmology. Follin had worked with Gamow, Alpher, and Herman at the Applied Physics Laboratory of The Johns Hopkins University and he was very mindful of the suggestion that it might be possible to obtain direct evidence of the Big Bang or its immediate sequels. "We are going to look for it," he told Sandage, "and find it!"

"I thought he was crazy," Sandage said to me later. Specifically Follin believed that, with rockets, it should be possible to observe the light emitted by hydrogen as, after the Big Bang, it first began forming into galaxies. (It would be light originally emitted at the ultraviolet wavelengths known as Lyman alpha, now highly red-shifted. As of this writing, that experiment has not yet been attempted.)

Following the original Penzias-Wilson observation, the existence of the microwave glow was confirmed, by them and others, at a number of wavelengths and in a variety of ways. While the extraordinary discovery that we can, in effect, still "see" the fireball from which the universe was born is now generally accepted, the proposal that quasars are residual bits of that fireball has not caught on. The true nature of the quasars therefore remains unresolved, but most astronomers continue to believe that in viewing those with the most extreme red shifts (some quasars have now been found receding at more than 90 per cent of the speed of light) they are looking far out toward the "edge" of the observable universe and the beginning of time.

8

The "Little Green Men"

In November 1967, Jocelyn Bell, like many students, seemed hopelessly behind in her work—half a kilometer behind. But unlike most students she and her thesis adviser, Antony Hewish (later to receive a Nobel Prize for it), were in the process of making a discovery of far-reaching consequences. Indeed, had their early suspicions proved correct, it could have been the most exciting in human history.

She was half a kilometer behind, for that was the length of chart paper with three-track recordings of radio signals waiting to be analyzed. And the recorders of the receiving system, which she herself had helped build near Cambridge, England, were pouring out more chart paper at thirty meters a day.

The purpose of the project was to identify quasars by the extent to which their radio emissions scintillated, or "twinkled." As noted earlier, such twinkling indicates that the source is very pointlike. Stars twinkle, but the nearer, brighter planets do not. And so, if radio sources scintillated, it was likely that they were very far away and probably quasars. Hewish had concluded that the scintillation of radio sources—which is much slower than star twinkling—was not, as in the case of stars, caused by turbulence in the atmosphere but by clouds of ionized gas moving rapidly out from the sun—the "solar wind."

The observing program was ambitious. A four-and-a-half-acre field was covered with 2,048 dipole antennas (each consisting of two aligned rods cut to fit the wavelength being observed). Jocelyn Bell, aged 24, had been responsible for wiring them into a co-ordinated receiving system. (Many graduate students such as herself are required to perform "slave labor" in addition to working on their doctoral theses.) By July 1967, all had been ready, and recording began. With the earth's rotation sweeping the field of view across the heavens it still took several days to scan the entire sky visible to the system, whereupon the scanning was repeated. In this way fluctuating sources that were man-made could be eliminated. Only if, week after week, the signals were observed coming from the same spot among the stars would they be considered quasar candidates.

Bell did the data analysis, as her interpretation of the recordings would be her doctoral thesis. In examining those made on August 6 she noticed what she later called "a bit of scruff"—a wavering signal about 1 centimeter long on the 120 meters of chart representing one full coverage of the sky. At the time when the "scruff" had been recorded it was midnight, with the antenna aimed directly away from the sun—the direction shadowed from the solar wind by the earth itself. Quasar scintillation therefore seemed unlikely. The observation was dismissed—and almost forgotten—as of local origin.

In September Bell saw it again. "I began to remember that I had seen this particular bit of scruff before, and from the same part of the sky," she said later. "It seemed to be keeping pace with 23 hours, 56 minutes—that is, with the rotation of the stars."

It was a similar argument that gave birth to radio astronomy. When Jansky first recorded radio noise from the sky in the early 1930s he thought it was of solar origin, and it was only after the sun, in subsequent months, changed its position relative to the stars that he realized the source had not moved with the sun but remained fixed among the constellations (whose passage overhead occurs every 23 hours and 56 minutes, rather than every 24 hours).

By the end of September the "bit of scruff" had shown up on six (though not all) occasions as the constellation Vulpecula, the Little Fox, passed through the antenna beam. Hewish thought it might be one of the stars that periodically erupts or "flares," so he and Bell decided to operate a fast

Jocelyn Bell Burnell, who detected the first pulsars, with her antennas.

recorder as it passed, observing its fluctuations in greater detail. For weeks it virtually disappeared. Then late in November, according to Hewish's account, "Miss Bell undramatically announced: 'It's back.' " A high-speed recording, made on November 28, showed clocklike pulses at a repetition rate of slightly more than one per second.

When Jocelyn phoned Hewish with this astonishing result, he replied: "Oh that settles it, it must be man-made." No rhythmic phenomena known to astronomers occurred at so fast a tempo—no spins, no orbits, no hypothetical vibrations. The shortest variable-star periods were about eight hours.

"We considered and eliminated radar reflected off the moon into our telescope, satellites in peculiar orbits, and anomalous effects caused by a large, corrugated metal building just to the south of the 4½ acre telescope," Bell said. Even though the source seemed fixed among the stars, Hewish believed it must be in the vicinity of the earth. "After all," he said later, "seasoned radio astronomers do not make the mistake of supposing that every queer signal on their records is truly celestial; in 99 cases out of 100 peculiar 'variable radio sources' turn out to be some kind of electrical interference—from a badly suppressed automobile ignition circuit, for example, or a faulty connection in a nearby refrigerator." Because of the association of the source with passage of the constellation Vulpecula, rather than with clock time, Hewish thought perhaps some other observatory was transmitting signals when that particular spot passed overhead. He consulted a number of astronomical colleagues, but none were doing so. "Still skeptical," he then arranged for high-precision time-signal ticks broadcast at 1-second intervals to be recorded with the incoming radio pulses. "To my astonishment," he said, the pulses proved so clocklike that their time-keeping was accurate to 1 part in 10 million—that is, to within 1 second in every 4 months. The pulse rate later proved remarkably steady at 1 every 1.33730113 seconds. Furthermore, when observed in a narrow-frequency range, the pulses themselves lasted only .016 second, implying—by the same argument applied to quasar outbursts—that the source must be very small.

"Having found no satisfactory terrestrial explanation for the pulses," said Hewish, "we now began to believe they could only be generated by some source far beyond the solar system, and the short duration of each pulse suggested that the radiator could not be larger than a small planet. We had to face the possibility that the signals were, indeed, generated on a planet circling some distant star, and that they were artificial.''

"Without doubt," Tony Hewish recounted later, "those weeks in December 1967 were the most exciting in my life." To announce that the beacon of another civilization had been detected would obviously create a worldwide sensation—and it might turn out, as in the case of the Soviet report on CTA 102, that a natural explanation would present itself.

"Just before Christmas," Bell recounted later, "I went to see Tony about something and walked into a high-level conference about how to present these results. We did not really believe that we had picked up signals from another civilization, but obviously the idea had crossed our minds and we had no proof that it was an entirely natural radio emission. It is an interesting problem—if one thinks one may have detected life elsewhere in the universe how does one announce the results responsibly? Who does one tell first? We did not solve the problem that afternoon, and I went home that evening very cross—here I was trying to get a Ph.D. out of a new technique and some silly lot of little green men had to choose my aerial and my frequency to communicate with us. However, fortified by some supper I returned to the lab that evening to do some more chart analysis.''

Shortly before the lab closed for the night she saw, in the record from a different part of the sky, another bit of "scruff." She checked earlier recordings from the same region and found that it had occurred before. The lab would soon be locked up, but at about 1 A.M. that patch of sky would pass over her antenna and she was there to observe. "It was a very cold night," she told an interviewer, "and the telescope doesn't perform very well in cold weather. I breathed hot air on it, I kicked and swore at it, and I got it to work for just five minutes. It was the right five minutes, and at the right setting. The source gave a train of pulses, but with a different period, of about 1¼ seconds.''

Soon a total of four such sources had been found, each with its characteristic pulse rate. In keeping with astronomical tradition they were given letter-number designations. Because of their suspected artificial nature they were listed as LGM-1 (the first discovered) through LGM-4, the letters standing for "little green men."

Before they announced their discovery, however, the Cambridge group had found that LGM-1 failed one test for artificiality. This was the variation in its pulse rate to be expected if the signals came from a planet orbiting another sun. Because the pulse rate was so very uniform, even by the standards of an atomic clock, variation in the observed rate should have shown up as the planet during its orbital flight varied its speed relative to the earth—a variant of the much-used Doppler shift. This effect could not be detected, even though that of the earth's own relative motion was evident. The possibility was not ruled out that a transmitting civilization had adjusted the pulse rate to keep it steady despite orbital motion by the transmitter. This would be done, for example, if—as one radio astronomer suggested—the pulses were serving as navigational aids for long-range space travel, much as loran beacons are used by ships and planes on earth. But as it seemed less and less likely that they were artificial, the term LGM was dropped and the sources came to be known as pulsars.

Because the higher-frequency part of each radio pulse arrived ahead of the lower-frequency part, it was possible to estimate the distances to each pulsar. Low-frequency waves are slowed by electrons in space more than high-frequency waves. Therefore calculating the distance to the point of origin was like figuring out how far two runners have traveled before the first crosses the finish line. If you know exactly how fast each of them ran and how much ahead the faster one was at the finish, it is easy to calculate how far they had to run for the winner to gain that much of a lead. While there was some uncertainty about the number of electrons adrift along the path of the radio pulses (and therefore of the slowing effect), rough distance estimates were possible, and it was clear that all the pulsars were within the Milky Way Galaxy.

But what were they? Hewish at first could think of no celestial phenomenon that might produce such rhythmic pulses. Without explaining why, he began spending a lot of time in the library of Cambridge University's optical observatory. His friends there, he said, "were surprised to see a radio astronomer taking so keen an interest in books on stellar evolution"—normally the province of optical astronomers.

He zeroed in on the manner in which stars evolve and finally collapse because, he thought, white dwarfs or neutron stars might be the answer. While the 1933 proposal of Zwicky and Baade that big stars, in their death throes, explosively collapse into neutron stars had almost been forgotten, the idea that such extremely dense objects might exist had recently been revived, notably to explain brilliant X-ray emissions from certain points in the sky. Alastair G. W. Cameron of NASA's Institute for Space Studies in New York had proposed that superdense stars would alternately expand and contract like a beating heart. A formula defining the rate of such oscillations in terms of the density and size of the star had been derived, but when Hewish tried applying it to white dwarfs it did not work. Such stars simply were not small enough and dense enough to oscillate at rates as fast as once a second.

When Hewish, in Stockholm, presented the lecture traditionally given by recipients of a Nobel Prize, he said discovery that one of the four original pulsars had a rate of only a quarter second "made explanations involving white dwarfs increasingly difficult." Neutron stars seemed a better candidate, although whether such objects really exist remained uncertain. In 1942 Baade had pointed to a star in the center of the Crab Nebula—that spectacular residue of the supernova seen in 1054—as possibly such an object. He picked it among several stars visible in that region because, unlike the others, it seemed to be moving in company with the nebula itself against the backdrop of distant objects. His suggestion was reinforced later by his colleague Rudolph Minkowski, although at the time there was little more on which to base the proposal except the unusual nature of the star: Its light seemed devoid of any spectral features at all.

When Hewish, with Jocelyn Bell and others of the Cambridge group, reported their discovery in the issue of *Nature* for February 24, 1968, they included the hypothetical neutron stars with white dwarfs as objects whose oscillations might produce the radio pulses. "If the suggested origin of the radiation is confirmed," they said prophetically, "further study

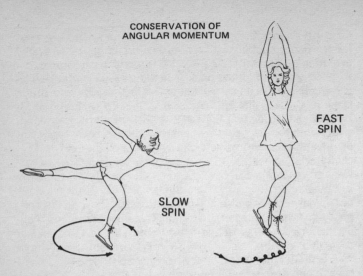

**CONSERVATION OF
ANGULAR MOMENTUM**

A figure skater can slow her spin rate by spreading her arms and increase it by bringing them close to her body. The principle, applicable to all spinning bodies, is known as the conservation of angular momentum.

may be expected to throw valuable light on the behaviour of compact stars and also on the properties of matter at high density.''

The idea that throbbing oscillations were to blame did not seem very plausible, and it was Thomas Gold of Cornell University who realized that spin was a much more likely explanation. Gold, who with Fred Hoyle and Hermann Bondi had helped formulate the ''steady state'' view of the universe, was noted for his highly original—and often controversial—ideas.

If, he said, pulsars are the hypothetical neutron stars, they should be spinning very fast. As a rotating star collapses, its spin rate must increase radically, unless it sheds much of its rotational energy—its ''angular momentum.'' This is often illustrated in terms of a figure skater whose spin rate increases when the arms, formerly spread, are brought close to the body. In the case of a neutron star the effect, Gold realized, would be extreme. For example, if a star like the sun, which spins once a month, collapsed to the assumed, ten-kilometer diameter of a neutron star, it would spin one thousand times a second! Actually a good deal of material would probably be thrown off during collapse, carrying with it enough angular momentum so that the final spin rate would not be so fast, but Gold estimated that a neutron star might start off spinning more than one hundred times a second. It would then slow gradually as it shed material and energy (including its pulses).

The magnetic field of the star would also become enormously concentrated. Such a field is often represented by force lines arching out into space. Those lines are, in effect, conserved during collapse, just as is the angular momentum, becoming tightly crowded together. Magnetic-field strength on the surface of a sun-sized star would increase ten billionfold if it collapsed to neutron-star size.

So powerful a magnetic field would clutch the ionized gas, or plasma, enveloping the neutron star with such force that a large volume of the plasma would spin with the star. This whirling envelope would extend out to where its outer edge was moving at close to the speed of light. There, at points of suitable magnetic alignment, plasma would be thrown off, generating a highly directional beam of radio waves (and, perhaps, light waves). This beam would sweep around the sky like an airport beacon. The resulting "flashes," as seen at the earth, would be highly rhythmic, but particularly with the youngest, fastest-spinning pulsars, the rate, according to Gold's hypothesis, should be slowing in an observable manner.

When the existence of pulsars became generally known, radio astronomers around the world scrapped other plans in order to begin observing them—and looking for more. Before the year (1968) was out, some two dozen had been discovered. By far the most exciting was one that seemed squarely in the middle of the Crab Nebula. It was first detected with the 300-foot dish at the National Radio Astronomy Observatory in Green Bank, West Virginia—an antenna whose aim can be adjusted only in a north-south arc, with dependence on the earth's spin to swing it from east to west. At first neither the pulsar's position nor its pulse rate were precisely known, but, as the observers, David H. Staelin and Edward C. Rei-

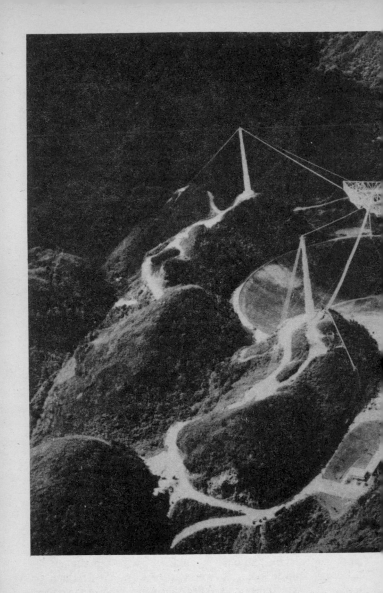

The world's largest radio astronomy antenna, 1,000 feet (305 meters) in diameter and with a surface area of twenty acres is suspended in a bowl-shaped valley near Arecibo, Puerto Rico. (*Arecibo Observatory, National Astronomy and Ionosphere Center*)

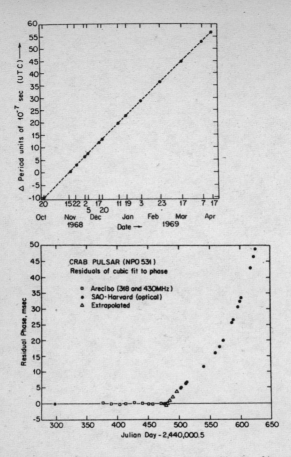

The steady slowing of the Crab Pulsar is illustrated in this plot of its rate during part of 1968 and 1969. It provided dramatic confirmation of the hypothesis that it is related to the spin of a neutron star gradually losing momentum. On the right the changes in rate are greatly exaggerated, showing that the slowdown is parabolic (rather than linear), as is typical of a top spinning down. There are also slight "glitches" or departures from uniformity as the star, with less angular momentum, collapses slightly in "starquakes." Although the shrinkage in the most severe quake is estimated at only ten millionths of a meter, the gravitational energy released equaled that radiated by the sun in a year. On earth it would be matched by an eight-kilometer subsidence of the earth's surface. (*Diagrams courtesy Arecibo Observatory, National Astronomy and Ionosphere Center*)

fenstein III, said, its association with the Crab Nebula, clearly a supernova remnant, "would support the view that pulsating radio sources may be neutron stars formed in explosions of supernovas." Further observations placed it near the center of the nebula, pulsing at the extremely fast rate of thirty times a second. This, in Gold's theory, was what one would expect from a pulsar born in the collapse of a star only 914 years earlier. The pulses themselves were so short—about three thousandths of a second—that they could only be generated within a very small area (again owing to the speed-of-light limitation). This conformed to the neutron-star concept, since such stars would be only a few kilometers in diameter.

Most impressive of all, observations with the thousand-foot dish, suspended in a bowl-shaped valley near Arecibo, Puerto Rico, showed that the pulse rate was slowing. It was doing so by only about a millionth of a second per month, but the effect was unmistakable. Careful observations showed, in fact, that virtually all pulsars are slowing very gradually. Gold's theory had received dramatic support on all counts.

Looking back on the flash of insight that led to his proposal, Gold said he was most impressed by "how fantastically stupid we really are when it comes to working out the consequences of something that we understand." All scientists, he said, knew "perfectly well" that when a rotating object with a magnetic field contracts, its spin rate increases and its magnetic field becomes stronger in a well-defined manner. But, he said, "it just didn't occur to anybody to work out the rotational energy that would be involved for a neutron star—or the strength of the magnetic field that would be trapped in a neutron star. Why not? Just because there was no observation that forced us to do it. . . . Why didn't someone go ahead and make these calculations? He could perhaps have predicted the existence of pulsars. We really have to have our noses shoved into things before we begin to think." His comment recalls the long delay in searching for the residue of the fireball flash from which the universe was born.

Attention now turned to the star near the center of the Crab Nebula that Baade had suggested might be the supernova remnant, as well as to the whole central region of the nebula. Might so young a pulsar be emitting visible light—be a star flashing at thirty times a second? Since the star and its nearest neighbors were all very dim, it seemed hopeless to record on-and-off flashes with photographic exposures of less than one thirtieth of a second, particularly since, from the duration of each radio pulse, it appeared likely that the star, each time, shone for only about three thousandths of a second.

In the hope of catching such on-off flashing, astronomers at the University of Arizona's Steward Observatory devised a special observing system dependent on "artificial synchronization." If, for example, you thought a distant light that seemed continuously on was really flashing twenty-four times a second (the rate at which motion-picture frames are displayed), you could verify this if you could blink your eyes at the same rate. When the blinking was in step with the "on" part of the cycle you would see the light on. When in step with the "off" part of the cycle no light would be visible. Furthermore, if a single flash was too dim to be seen, but your eye could store the tiny bit of light that did come through, then after repeated flashes enough light might accumulate to be visible.

The plan was to use the Steward Observatory's 36-inch telescope atop Kitt Peak in southern Arizona. One of the observatory team, Donald J. Taylor, had developed a computer-controlled recording system that could be "tuned" to the assumed repetition rate of the pulsar and display each sweep of that time interval on a scope. The amount of light observed at each moment during the sweep was shown as a series of green dots across the scope. If no light was observed, the dots formed a more or less straight line along the edge of the scope, but the display was cumulative so that if, on successive sweeps, a tiny bit of light was repeatedly recorded at some particular part of the cycle, the dots there would begin to climb higher.

A special feature of this experiment was that voices of the participants were inadvertently tape recorded during the observations and their accounts were supplemented by later interviews (conducted by the Center for History of Physics of the American Institute of Physics). After the apparatus had been installed, tested, and set to match exactly the radio pulse rate, the telescope was aimed at the star Baade had identified.

In the scope a diaphragm with a tiny hole cut out all light except from an area twenty-two arc seconds wide (this, it will be recalled, is roughly the observed size of Mars when close to the earth). Light coming through the hole fell on a photomultiplier—a device that could amplify a light flash so weak only a few photons (light particles) were received. It was the output of this device that was chopped into segments and superimposed on the display scope.

When all was ready the apparatus was turned on and the three experimenters—Don Taylor, John Cocke, and Michael Disney—began watching the display of dots. "What we were looking for," said Disney, "was that several of these dots should race out in front of the others, because this would tell us it was a pulse of light coming from this pulsar. . . . Well, for a few moments we were quite excited because we really thought that Baade's star would be the place to find this pulsar. And, in fact, there was no sign of a pulsar at all, so we were all a little bit let down."

Taylor, the electronics wizard who had developed the observing system, had to drive down the mountain and back to the university in Tucson, but the other two tried again the next night. "We repeated essentially what we had done the previous night," said Disney, "and we didn't find any pulsar at all."

As Cocke put it: "I do remember feeling pretty discouraged about the whole thing. Of course, we weren't very optimistic to begin with—or at least, I wasn't. And so I wasn't feeling as discouraged as I would have if I'd had really high hopes."

They had allocated two more nights for the experiment, but both were overcast. "The next two days," said Cocke, "were spent more or less walking around the mountain under the clouds, trying to think about what to do next." He decided to rework the calculations that had gone into the setting up of the experiment and suddenly realized a subtle factor had been overlooked: They had not properly corrected the pulse rate for its alteration by motion of the observatory toward or away from the source. While the pulse rate was known very precisely, it would seem faster when the observatory (because of the earth's spin and orbital motion) was approaching the source and slower when the opposite was true.

"I felt like a real idiot," said Cocke. "And I was a bit dumbfounded, because now, having gone through all that time to make these observations, now I was faced with the prospect of having to go back and do the whole darned thing over again—again with the feeling that it was utterly futile anyway, and that, really, the whole thing was a waste of time."

On the night of January 15, 1969, Cocke and Disney were back at the observatory with the corrections made. They started the apparatus, which ran automatically, then crouched in front of the little scope and began watching the line of green dots. Bob McCallister, the observatory's night assistant, turned on the tape recorder in case anyone wished to make comments for the record, and he spoke into it:

"This next observation will be observation No. 18." He then forgot to turn off the recorder and so what followed became a verbatim part of scientific history. On the tape one can hear the hum of equipment and the reverberation of sharp sounds within the observatory dome. Then Disney's voice (his choice of language and accent revealing his British origins) remarked rather calmly:

"We've got a bleeding pulse here." Almost immediately the green dots near the center line of the scope had begun to climb.

"Hey," said Cocke in amazement. After a long pause: "Wow! You don't suppose that's really it, do you? Can't be."

The dots representing light intensity were clearly building up a peak partway through the pulse period.

"It's right bang in the middle of the period," said Disney. "Look. I mean, right bang in the middle of the scale. . . . It's growing, too. It's growing up the side a bit, too!"

"God! It is, isn't it?" said Cocke, and Disney cut in: "Good God, you know that looks like a bleeding pulse," whereupon both men giggled slightly hysterically.

"It's growing, John. Look!" cried Disney. ". . . It's really building up. Look at that. . . . There's not one left behind now. See, look, not one of those dots left behind."

"By God—yeah—uh huh," said Cocke.

"There's a second something over here," said Disney as a weaker intermediate pulse became evident.

"Well, we expect two," said Cocke, "a small pulse and a larger pulse, remember?"

"Uh huh, right," said Disney. Then, later: "I won't believe it—I won't believe it until we get a second one."

"I won't believe it until we get a second one and until the thing has shifted somewhere else," said Cocke. They were worried that appearance of the pulse exactly in the middle of the scope represented a defect in the electronics rather than a pulsar. But if in a second recording the pulse reappeared and at a different position on the scope, that would be persuasive evidence it was the real thing.

"God, just come and look at it down here," said Disney, and they broke into laughter again. "This is a historic moment!"

"Hmm," added Cocke in a softly spoken footnote, "I hope it's a historic moment."

One test of the authenticity of the signal would be to change the aim of the telescope so that it no longer pointed at the suspect star and see whether or not the effect still showed up.

"Let's move off that position and move somewhere else, and see if we get the same thing. All right?" said Cocke. "I hope to God this isn't some sort of artifact in the instrumentation."

They aimed the telescope to one side and the dots on their screen climbed to a peak—much smaller than before, but it seemed to show the phenomenon was in their instruments, rather than in that star far out in the Crab Nebula, and their hearts sank.

Another test was to change the pacing of the scanner so that it no longer matched that of the pulsar. This killed the effect entirely, which revived their hopes. Perhaps, they thought, they had not swung the telescope far enough off the star and some of its light was still entering the pinhole. They moved it well to one side and this time there was no evidence of pulsing at all. Finally they returned the aim to the pulsar and set the proper timing for a repetition of the first observation. McCallister started the recorder again.

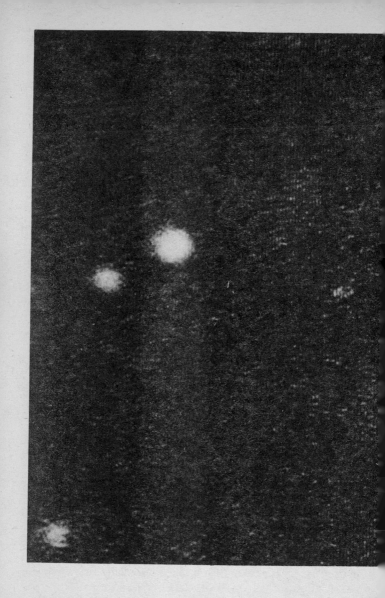

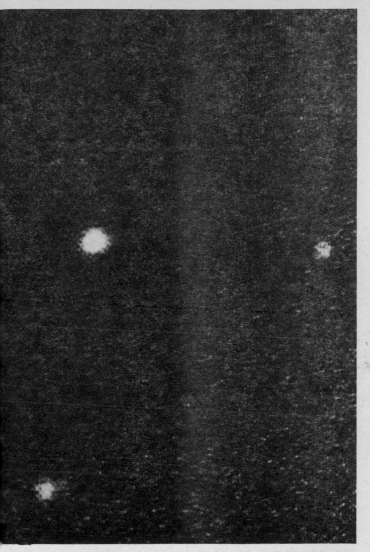

Using an exposure system timed to synchronize with the flashes of the Crab Pulsar, the Lick Observatory obtained these photographs of the star "switched on" (left) and "switched off." (*Lick Observatory photograph*)

Synchronous optical scanning of "Baade's Star" in the Crab Nebula, timed to the thirty-per-second rate of the radio pulsar, produced evidence of a flash in the midst of each sweep, as shown in this record made by the University of Arizona astronomers on Kitt Peak. On the left is evidence for a second pulse. (*University of Arizona*)

"This is observation No. 23," he said into it. "It's a repeat of observation 18 of this night."

Disney mumbled something about a "small prayer" and then cried out: "Here she comes." The peak of little dots began to grow. "It's there, all right," said Cocke. "Look at it. God!"

"Well, that's just about—there couldn't be anything more definite than that, could tha'?" said Disney.

"Come on, tape, hold out," said McCallister.

"Are we running out of tape?" asked Cocke.

"Yeah," replied McCallister.

"Any room for the American National Anthem on that?" said Disney. They laughed and a moment later the tape ended.

On four nights they detected the optical pulsations in this manner. With the position of the source better known, a smaller diaphragm was used (its hole five seconds wide). Whereas five thousand pulses had to be superimposed to achieve a clear effect with the larger hole, the small one, when aimed at the star identified by Baade, produced a much stronger signal from only three hundred pulses.

News of the discovery was sent to other observatories by the Central Bureau for Astronomical Telegrams, operated on behalf of astronomers around the world by the Smithsonian Astrophysical Observatory in Cambridge, Massachusetts.

Within days the finding had been confirmed by two other observatories and eventually it was possible for the Lick Observatory, with its three-meter telescope and a photographic counterpart of artificial synchronization, to obtain photographs showing that area of the sky with the star clearly visible and then with no star in sight at all.

The discovery of pulsars, more than three hundred of which are now known, has introduced a whole new field of research, for, as Hewish and his colleagues predicted, it has made possible the observation—although at very long distance—of the material forming atomic nuclei under conditions beyond the reach of any laboratory. Probably no single object outside the solar system has received so much attention in recent years as the Crab Nebula and its pulsar. Its closest rival is the only other pulsar with a high pulse rate. It lies in the constellation Vela, the Sails (being the sails of the "ship" forming the superconstellation Argo), and emits one radio pulse every tenth of a second. Like the Crab, its pulses seem to span much of the electromagnetic spectrum from the very short wavelengths of X rays through those of visible light to radio waves. Because X rays cannot penetrate the earth's atmosphere, the pulsations at those wavelengths have been detected with rocket shots, balloons drifting near the top of the atmosphere, and earth satellites.

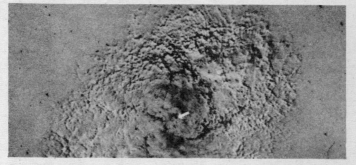

Superimposing two photographs of the Crab Nebula taken fourteen years apart shows its expansion. A 1964 photograph taken by Guido Munch is superimposed (in negative) on a 1950 one by Walter Baade (in positive). Thus the earlier position of each feature is shown in white and its later one in black. "Baade's Star" (the pulsar) is indicated by an arrow. (*Hale Observatories*)

In January 1977, the Anglo-Australian Observatory in Siding Spring, Australia, finally was able to record Vela optical pulses after the tracking station of the American space program, near the Australian capital city of Canberra, had provided a pulse rate accurate to a tiny fraction of a second and Australian radio astronomers had obtained a more precise position.

In generating pulses that span so wide a range of wavelengths these young pulsars are indeed remarkable "lighthouses in the sky." If the awesomely beautiful cloud of filaments forming the Crab Nebula is the remnant of the supernova in which its pulsar was born, where is the remnant of the supernova that left the Vela pulsar? Because the pulse rate of the latter is slower, the explosion presumably occurred many centuries earlier and its debris is far more widely dispersed.

In 1952 an Australian, Colin S. Gum, identified a vast and very faint nebula spread across the southern sky. It spans sixty degrees of the heavens there—so huge and faint that during all the centuries since telescopic observations began it passed unnoticed. Known for its discoverer as the Gum Nebula, it is some three thousand light-years wide. Its glow was originally attributed, as in many other such clouds, to illumination (more precisely, atomic excitation) by radiation from hot stars within it. In 1971, however, four scientists—three of them associated with the Goddard Space Flight Center in Maryland and one with the Kitt Peak National Observatory—proposed that the Gum Nebula is, in fact, the remnant of the great explosion that produced the Vela pulsar.

The Vela and Crab pulsars, spinning at almost incredible speed, are subject to sudden changes of rate or "glitches." There may be "starquakes" that occur when the neutron core suddenly collapses to greater density as the spin rate (and therefore the centrifugal effect) are reduced, or when some other internal change takes place. Another explanation would be material falling onto or thrown off the fast-spinning object.

Some physicists now look to pulsars as the answer to a long-standing mystery: the origin of the highest-energy cosmic rays. The "rays" consist primarily of atomic nuclei traveling at almost the speed of light, some of them with sufficient energy to penetrate into tunnels and mines deep underground. While various ways have been proposed for the acceleration of these particles (the lowest-energy components are generated by stars like the sun), those of very high energy had not been satisfactorily explained until the discovery of pulsars. As pointed out by Tom Gold, Jeremiah P. Ostriker at Princeton University, and others, the process producing pulses from very young, fast-spinning pulsars, like the Crab and Vela, could throw off extremely high-energy particles.

While the pulsar discovery was one of the most dramatic developments of twentieth-century astronomy, it was also marked by controversy, following the award of a Nobel Prize to Hewish in 1974. He shared the prize with Sir Martin Ryle, leader of the Cambridge radio astronomy group, who was honored for innovations in antenna design that had made possible the detailed mapping of radio sources. Hewish was cited "for his decisive role in the discovery of pulsars," but a few months later, Hoyle—by then Sir Fred Hoyle—protested that Jocelyn Bell, now Mrs. Burnell, should also have shared in the award.

"There has been a tendency to misunderstand the magnitude of Miss Bell's achievement," said Sir Fred in a letter to *The Times* of London, "because it sounds so simple—just to search and search through a great mass of records. The achievement came from a willingness to contemplate as a serious possibility a phenomenon that all past experience suggested was impossible." In particular, he said, it was her recognition that the source was fixed among the stars that was critical. "Once this step had been taken," he wrote, "nothing that happened from then on could have made any difference to the eventual outcome."

Mrs. Burnell commented that such criticism of the award was "a bit preposterous." Actually, Nobel Prizes are often given for work in which many were involved and, as a rule, the awards committee does its homework carefully.

Now that the neutron-star speculation of Zwicky and Baade had been confirmed to the satisfaction of most scientists, many began to wonder if the even more extreme objects—black holes—might also be a reality. Supporting evidence came from an ingenious series of observations made from above the earth's atmosphere.

9

Uhuru and the X-ray Sky

Like lobsters and crabs on the ocean floor, little aware of what is going on beyond the waters above them, we live under a sea of air. But in the latter decades of the twentieth century we have lifted our eyes and our instruments above that sea and glimpsed some of the wonders beyond.

The atmosphere impairs our vision because it absorbs much of the "light"—that is, the electromagnetic radiation—reaching the earth from afar. Our eyes are adapted to see in a very limited part of that radiation—those wavelengths embracing the rainbow colors—because such waves readily pass through air and are thus useful to us. This is not true beyond the red (or longer wavelength) end of the visible spectrum—in the infrared and in much of the radio spectrum. Nor is it true in the other direction—in the ultraviolet and X-ray wavelengths.

The blame does not lie entirely with the atmosphere. For example, in the far reaches of space between the stars hydrogen atoms also absorb some of the ultraviolet radiation, and clouds of dust further limit our vision out to great distances. The possibility that, by climbing into the sky and looking at the universe through new wavelength windows, we might make exciting discoveries had been encouraged by the new science of radio astronomy and by the arrival in the United States of captured German V-2 rockets after World War II. They were called V-2s because they were the second of Germany's "retaliation weapons" (*Vergeltungswaffen*), the V-1 having been a "buzz bomb" that was essentially a pilotless aircraft. The V-2 missiles were the first to reach supersonic speed, and more than thirteen hundred had been fired at Britain as well as a large number at Belgium.

They were being made in what was then the world's largest underground factory, at Nordhausen, which was in the zone that, by prior agreement, had been allocated for occupation to the Soviet Army. The United States Army, however, managed to get there first and brought out enough parts to assemble about a hundred rockets. They were shipped to the White Sands Missile Range in New Mexico to be used for both military and scientific research.

The V-2 was fourteen meters tall with a rocket nozzle at its base so large a man could crawl through it. When fired vertically it could soar more than one hundred kilometers. One of the first goals was to learn the nature of sunlight beyond the violet end of the spectrum (in ultraviolet and X-ray wavelengths). Radio communications past the horizon are made possible (at appropriate wavelengths) because the waves are bent back toward the earth when they reach certain levels of the upper atmosphere. One reason, it was suspected, is that some form of "light" from the sun (probably ultraviolet generated by hot hydrogen) frees electrons from atmospheric atoms (the process called ionization) and these electrons then bend the radio waves.

The true nature of these unseen components of sunlight, however, was not known. The V-2 offered scientists a chance to have a look, and a group at the Naval Research Laboratory (across the Potomac River from Washington's National Airport) decided to take advantage of it. A detector (spectrograph) sensitive to the ultraviolet wavelengths under suspicion was mounted in the nose of a V-2 and the rocket was ignited on June 28, 1946—less than a year after the war's end.

"The result," reported Herbert Friedman of the Naval Research Laboratory, "was catastrophic." Instead of soaring into the sky, Friedman said, "the rocket returned to earth, nose down, in streamlined flight and buried itself in an enormous crater some eighty feet in diameter and thirty feet deep. Several weeks of digging recovered just a small heap of unidentifiable debris; it was as if the rocket had vaporized on impact."

This V-2 launch on May 10, 1946, looked good to begin with. (*U.S. Air Force Photo*)

In the early days of rocket experiments loading instruments into the nose of a V-2 missile was a risky business. Not until later did gantries make the job easier and safer. (*R. Tousey, Naval Research Laboratory*)

These early shots, flown by a variety of investigators as well as the Navy group, provided their onlookers with a variety of unexpected thrills. As Friedman recalled later, "Some V-2 rockets somersaulted end over end; some exploded on take-off; and one landed on the edge of Juarez, just over the Mexican border." All of the first five attempts ended in disastrous nose dives. Fortunately none hit the blockhouse where the scientists and rocket engineers were sheltered, but when one exploded on the pad and fell toward the onlookers, Friedman said, it "burned so furiously that smoke and flames seemed to engulf the blockhouse—a very scary experience for those inside."

On later flights, to slow the descent of the instrument package to the "comparatively gentle" speed of 200 miles an hour, explosive bolts blew off the nose and tail. These parts then fluttered down and their heavy metal cassettes of recording film could be recovered intact.

As suspected, ultraviolet wavelengths proved chiefly responsible for daytime ionization of the upper atmosphere, but it was suspected that solar X rays might also play a role. Film sensitive to such rays, flown on a Navy rocket in 1948, came back blackened. Friedman devised more precise measurements that, on a V-2 shot the next year, confirmed that the sun shines weakly in X rays.

Also of special interest was finding out what causes radio blackouts when the flash of an eruption on the sun, or solar "flare," reaches the earth. Might a burst of X rays ionize the upper air enough to absorb the radio transmissions? The challenge was to spot a flare and get instruments into space minutes later—before the flare subsided. James A. Van Allen, later to discover the earth-encircling radiation belts that bear his name and then at the Applied Physics Laboratory of The Johns Hopkins University, came up with an ingenious scheme. Before the supply of V-2s was exhausted he had proposed a less expensive rocket which, developed by the Navy as the Aerobee, had become the prime workhorse of rocket research, but once its liquid fuel had been loaded under gas pressure it had to be fired within an hour or two. It could not be held in readiness for days, awaiting a flare.

Van Allen's answer was to hang small Deacon rockets from balloons that lifted them high in the atmosphere. When a flare was sighted the rocket could be fired by radio command, carrying instruments even higher. This balloon-rocket combination, called a Rockoon, could be long held in readiness since the Deacon was solid-fueled. Rockoon shots showed that X rays are to blame for the blackouts, as did subsequent experiments where the small rocket was lifted by a Nike antiaircraft missile rather than a balloon.

While one Rockoon had recorded very energetic ("hard") X rays, even though no flare was occurring on the sun, it seemed unlikely that the rays had come from far beyond the solar system. Even the most intense solar X-ray bursts were far too weak to be detectable across the great void between stars, and there was no reason to believe other stars would be very much brighter than the sun at such wavelengths. Furthermore, attempts by rockets to record low-energy ("soft") X rays from the night sky had revealed none detectable with the instruments in use.

The discovery that, if one scans the sky at X-ray wavelengths, there is, in fact, a great deal to "see"—including the first evidence for black holes—was, to some extent, accidental, the chief target of the experiment being the moon. It was conducted by a group of physicists from Cambridge, Massachusetts, organized into a company called American Science and Engineering. It had been formed chiefly to conduct secret research on the effects of nuclear-weapons explosions for the Department of Defense. The chairman of the board was Bruno Rossi, a lean, aristocratic-looking physicist who had left Italy in the Fascist period to become a professor at the Massachusetts Institute of Technology. He was an authority on those high-energy particles from space known as cosmic rays.

Ever since children were first told the moon was made of green cheese, scientists had wondered about its true composition. The discovery that the sun shines in X rays as well as in visible light suggested a way to find out. Solar X rays (or bombardment by high-velocity gas blowing out from the sun), it was reasoned, should make the lunar surface reradiate X rays (that is, fluoresce) at wavelengths indicating the nature of its materials.

To conduct such research American Science and Engineering (AS & E) hired a twenty-eight-year-old physicist, Riccardo Giacconi, who three years earlier had come over as a Fulbright scholar from Italy, where he had studied under a student of Rossi. The company, in requesting Air Force support for an attempt by rocket to record such emissions, noted that objects other than the sun and moon, such as supernova remnants or flare stars, might emit X rays, but that it seemed unlikely they could easily be observed at such great distances. The prime target was therefore to be the moon.

After one failure with a smaller rocket, an Aerobee was launched from White Sands on June 12, 1962, and soared to 230 kilometers. For five minutes, fifty seconds it was above virtually all the atmosphere, sweeping its X-ray eyes across the sky. If any single shot into space marked the birth of X-ray astronomy, this was it. Around the nose of the rocket were three Geiger counters covered by mica windows coated with lampblack to prevent penetration by radiation other than X rays. An electronic eye looked out the side of the rocket to provide information on the latter's aim so that the Geiger counters' fields of view could be calculated for each moment of the flight.

The moon was one day past full and thirty-five degrees above the horizon, twenty degrees east of south. From the rate at which the electronic eye saw the moon it was clear the rocket was spinning twice a second and was aimed almost straight up throughout its soaring flight.

As soon as it neared the top of the atmosphere the Geiger counters began to record radiation whose intensity varied during each spin of the rocket, reaching a peak every time a properly functioning counter (one had failed) was aimed about fifteen degrees west of south. This peak showed up much more strongly in the counter that had the thinnest mica window, implying that the X rays were relatively "soft." No matter where the counters were aimed, the readings never dropped near zero, which indicated that the whole sky was aglow with diffuse X rays—so much so that none could be detected from the moon. In fact, later shots revealed the moon as a "hole" in the observed X-ray sky.

Because the force lines of the earth's magnetic field were oriented almost exactly in the direction from which the most intense radiation was observed, the AS & E scientists were worried that the source, instead of being distant, might be an effect caused by high-energy particles spiraling along those field lines. Such trapped particles form the Van Allen belts, which envelop the earth.

There was, however, a hint of a second X-ray source in the general direction of the two strongest emitters of radio waves beyond the solar system—Cassiopeia A (a supernova remnant) and Cygnus A (an "exploding" galaxy). The main peak was toward the core of our own Milky Way Galaxy. On this and other grounds Giacconi and his colleagues (Herbert Gursky, Frank R. Paolini, and Bruno Rossi) decided that they had for the first time detected X-ray sources beyond the solar system.

To verify so important a discovery AS & E launched two more shots—in October 1962 and June 1963. The sources proved to be in the same positions among the stars, despite the earth's orbital movement around the sun, so it was clear they were not a local effect. There were also hints of X-ray emission from other parts of the sky, and it appeared that the most powerful source was not precisely in the direction of the galaxy's center.

To learn the location and nature of the sources was no easy task, for one cannot use ordinary telescopes or other optical devices when X rays are being observed. Such rays are not focused by a lens as are the longer-wave forms of light and they are reflected by mirrors only at very shallow grazing angles. One way to find their approximate locations among the stars was to aim detectors at the sky through tubes that limited the passage of X rays unless they shone directly down the tube. At least one ultraviolet shot used the tubes of hypodermic needles for this purpose.

Herbert Friedman's group from the Naval Research Laboratory aimed its X-ray detectors through hexagonal tubes, clustered like the cells of a honeycomb. Such an arrangement, with more than six times as much detector area as the AS & E experiment, was lifted by an Aerobee on April 29, 1963, and narrowed the location of the more powerful of the two sources that had been discovered. Eight times, as the rocket spun and wobbled, the detectors swept across the source, and these strong emissions were clearly coming from Scorpio—the constellation, with its scorpionlike tail, that, for those in northern latitudes, spans the southern sky on summer evenings. Even at the summit of the flight the core of the galaxy was below the horizon, so it obviously was not the origin.

Another source, with one eighth the power, was seen near the Crab Nebula, and Friedman proposed that both objects were neutron stars left by former supernovas. That such stars would "shine" in X rays had been postulated by several theorists. As noted earlier, it was these rocket observations that revived the neutron-star idea thirty years after Zwicky and Baade had proposed their existence.

An opportunity to see whether the X-ray emissions were, in fact, coming from a tiny point in the center of the Crab (the pulsar there had not yet been discovered) occurred on July 7, 1964, when the moon was to pass across the nebula, progressively cutting off ("occulting") its emissions. Friedman's group hoped to use the moon much as had Hazard two years earlier, when he located the first quasar with his Australian radio telescope. In this case, however, the observation could not be made from the ground. Furthermore, the rocket could only remain above the atmosphere five of the twelve minutes when the moon would be passing across the nebula.

It was decided to time the shot to record the central part of this eclipse. The Aerobee waiting on the pad had been fitted with a new control system that had failed on all six previous launch attempts but, Friedman reported, "for the very special occultation of the Crab, happening only once in nine years, it performed perfectly." From the way in which the X rays were gradually cut off and then gradually reappeared it seemed that most of the emissions were from the nebula as a whole. "Clearly," Friedman wrote, "the Crab X rays are not generated in a neutron star." (Not until later was it established that there is, in fact, a neutron star at the center of the nebula pulsing thirty times a second across the entire spectrum from radio waves to visible light, X rays, and gamma rays.)

Riccardo Giacconi, leader of the Navy group's gentlemanly rivals at AS & E, pointed out that Friedman's "very beautiful experiment," making use of the moon, "succeeded for the first time in identifying an X-ray source with a previously known celestial object." Since the distance to the Crab was well established, it became possible to estimate the intrinsic brightness of its emissions at X-ray wavelengths. The findings were hard to explain.

Thus when Friedman and his Navy colleagues again sent X-ray detectors aloft in an Aerobee five years later—after discovery of the Crab pulsar—they found that it radiates ten thousand times more energy in its X-ray pulses than at the radio wavelengths first detected. In fact, the total energy production of the nebula at all wavelengths is one hundred thousand times that of the sun. How, astrophysicists wondered, could this radiation still be so intense more than nine hundred years after a supernova gave birth to the spinning neutron star and the expanding nebula around it?

The answer became clear when observations with the giant Arecibo dish, suspended in a bowl-shaped valley in Puerto Rico, recorded a very slight but steady slowing of the Crab's pulse rate—fifteen millionths of a second per year. When news of this reached Thomas Gold at Cornell he immdiately put through a phone call to Friedman at the Naval Research Laboratory. Gold, as noted earlier, suspected that a pulsar would gradually shed its spinning energy and slow down. Was it possible, he wondered, that the angular momentum shed by the Crab pulsar in its slowing was sufficient to account for the pulsar's radiated energy? What, asked Gold, was the total energy radiated by the pulsar in X rays? "At my answer," Friedman reported later (in his book *The Amazing Universe*), "he exclaimed with excitement that the fit was perfect! The energy released by the slowdown in the spin of the pulsar was an incredible ten thousand trillion trillion kilowatts. The transformation of that tremendous energy into radiation could account for all of the star's radio, visible, and X-ray power!"

The answer to the riddle of the Crab, Friedman said, is thus "a story of cosmic death and transfiguration." A star that had been radiating light in the ordinary manner collapsed, flaring up so that on earth it was observable day and night, to form a superdense pulsar whose angular momentum—spinning many times a second—enabled it to shed even more energy than it presumably did as a normal star.

Just as the discovery of quasars started a scramble to find more, those with access to rockets in the mid-1960s sent a succession of X-ray detectors into the sky. Improvements were made in aiming, in star sensors, horizon sensors, and magnetometers used to determine where the detectors were actually pointed, as well as in sensitivity of instruments and in ways to slow rocket roll or wobble for long exposures. By 1966 the Naval Research Laboratory, the team from AS & E and the Massachusetts Institute of Technology, and scientists from The Lockheed Missiles and Space Company had found about twenty sources. Most were in or near the Milky Way, indicating they were probably within our own galaxy. Exceptions seemed to include several of the "exploding" galaxies whose output in radio energy, considering their very great distance, had proved so astonishing. In 1964 the Navy scientists rated the source in Cygnus, now called Cyg X-1 (meaning Cygnus X-ray source No. 1), the second most powerful, but when they had another peek above the atmosphere a year later they found that it had dimmed to one quarter its earlier brightness, implying an enormous change in energy output—the first hint that there was something very remarkable about this object.

The key to discovering the nature of the newly detected X-ray sources was narrowing down their locations so astronomers with optical and radio telescopes could see what was there. Attention focused on the source that had been first detected, being (at least in lower-energy X rays) the brightest such source in the sky (the Crab was brighter at the shorter, more energetic X-ray wavelengths).

By now it was clear that the original observation had recorded the combined emissions of at least two sources. The strongest, being in the constellation Scorpius, was designated Sco X-1. To provide astronomers with a precise position Minoru Oda of the University of Tokyo, working with the AS & E group, devised a method that was ingenious and beautifully simple. It was a scanning system that could be likened to the window of a dungeon that forms a short tunnel through the wall with two sets of bars, one at the outer end and one at the inner end. The bars of both grids are closely spaced and parallel to one another.

If a brilliant light of small angular width passes by outside the dungeon, it will appear to those inside as a series of flashes that occur when the light can pass through the slits between both sets of bars. If the light source is broad, like the sun, this effect will be greatly diminished. The light intensity may vary, but only a little, since the sun is so wide that some light always gets through. If the source is narrow, the flashes will be most prolonged when the source is directly opposite the window.

In Oda's device, since tiny sources of X rays were being sought, everything was on a small scale, and grids of fine wires instead of iron bars were used. Because the wires and slits between them recorded lines, rather than points, to pinpoint the source required two separate units with their wire grids at right angles to one another. The system imposed no limit on detector area—an advantage in seeking to pick up weak emissions. The only factor limiting the width of the X-ray "windows" was the size of the rocket.

This system was carried into space on several flights, including one on March 8, 1966, aimed at Sco X-1. The rocket's guidance system was programed to keep it pointed toward that target, but because of the rocket's gentle rock and roll, the field of view of the detectors swept back and forth across the source. At the same time a camera, once every second, photographed the star field under observation, so the aim, at any moment, could be reconstructed.

Sco X-1 was in view of one or both of the detectors for fifty-five seconds, the readings being continuously radioed to earth. Then, as the rocket fell, the instrument package including the film was cast loose and parachuted safely to the ground some forty miles downrange from the White Sands launch pad.

Analysis of the results showed Sco X-1 to be no more than twenty seconds of arc wide—that is, starlike. Its location was narrowed to a row of rectangles each only a few seconds of arc on a side. Two of the rectangles seemed by far the most likely candidates. Oda communicated the information to colleagues at Tokyo Observatory along with the prediction of the AS & E group that, if visible, the object should appear as a star of the thirteenth magnitude (rather dim)—an estimate derived from the observation that, at longer X-ray wavelengths, it became increasingly dim and therefore in invisible light should be even weaker. Likewise it should shine most strongly in the ultraviolet, they said.

On a succession of nights from June 17 to 23 Tokyo Observatory's 188-centimeter reflector and a smaller telescope were trained on the two rectangles. Although, as the astronomers reported, the observations were "frequently interrupted by clouds, which are prevalent in Japan during this rainy season," they found, close to one of the rectangles, "an intense ultraviolet object" of the proposed magnitude. Its colors, they said, "are definitely peculiar and in the range predicted by the working hypothesis."

The results were cabled to Giacconi, and he immediately phoned Allan Sandage in California, who aimed the five-meter instrument on Mount Palomar at the object that same night. He not only confirmed the findings but also found it to be flickering in an extraordinary manner. During forty-two minutes of steady observation its brightness varied about 2 per cent every few minutes. Between July 17 and 18 its brightness through a B (blue) filter increased more than two and a half times. An examination of photographic plates at the Harvard College Observatory dating back to 1896 showed that it had been variable for a long time. While it bore some resemblance to a star that occasionally flares up—a nova—there was no sign in those plates of its having done so. "The most striking characteristic of the object," reported Sandage, the Japanese, and the Cambridge scientists in a joint paper, "is that it emits X rays in copious quantity." The energy emitted in that part of the spectrum, they said, "is about one thousand times greater than that emitted in visible light."

While the Tokyo and Palomar observations were being made, two astronomers atop Kitt Peak in Arizona (Hugh M. Johnson of Lockheed and C. B. Stephenson of the Case Institute of Technology) were conducting their own search for Sco X-1. While they had only the earlier rough positions, they zeroed in on a star that seemed highly variable, very ultraviolet, and of roughly the predicted magnitude. It was, in fact, the same object identified by the others.

Eight days after the March 8 rocket flight that enabled Tokyo and Palomar to identify Sco X-1, Neil A. Armstrong, who was to become the first man to tread on the moon, and David R. Scott orbited the earth seven times in a Gemini spacecraft, gazing at the universe in its seemingly black, star-studded beauty, free from the filtering effects of the atmosphere. A number of astronomers wondered what they might have seen had their eyes been adapted to X-ray wavelengths. It had begun to seem likely that the sky was peppered with "X-ray stars" manifesting phenomena fundamentally new to the sky-gazers.

Particularly tantalizing was the evidence of rapid and radical changes occurring in some objects, such as Sco X-1 and Cyg X-1. The rockets kept instruments above the atmosphere only about five minutes, so the observations were brief and their coverage limited. More extended observations were made with balloons that sometimes lifted instruments higher than 45 kilometers, above 99.9 per cent of the atmosphere. An MIT group led by Walter H. C. Lewin flew several balloons from Australia, where such sources as Scorpius would be nearly overhead. One in 1972 was the largest balloon ever flown up to that time, remaining aloft twenty-seven hours. But clearly the only way to survey the heavens systematically was from orbit, and an instrument was prepared to be carried on the first manned flight of Project Apollo—a test in earth orbit. It was an ill-fated mission. On January 27, 1967, during a dress rehearsal on the pad, a fire in the command module killed all three astronauts of the crew. It was a severe setback to the moon-landing program. No manned flight was attempted until twenty-one months later, and the X-ray device never made the passenger list.

Meanwhile Giacconi had proposed to NASA (the National Aeronautics and Space Administration) the development of a small, relatively inexpensive satellite to map the sky in X rays. It was to be one of the Explorer series of unmanned spacecraft and would also be designated Small Astronomy Satellite 1 (SAS-1). Probably no vehicle launched into space will ever again prove so productive relative to its modest cost. Of the $13.25 million allocated to the project, roughly $7 million was for the spacecraft, $5 million for the experimental payload, $1 million for the four-stage launching vehicle (a Scout rocket), and $250,000 for payment to the Italians who launched it. The Applied Physics Laboratory of Johns Hopkins built the spacecraft, AS & E developed the instruments, and NASA's Goddard Space Flight Center near Washington, D.C., undertook direction of the project with

Marjorie Townsend, a forty-year-old engineer, as its manager (the first time a woman had performed that role).

An essential element of the plan was placing the vehicle in orbit over the equator so that its detectors would be able to scan virtually the entire sphere of the heavens. The area that they swept on each rotation of the spacecraft would be a narrow band, but by altering the spin axis the areas covered could be changed to provide a systematic survey of the whole sky.

A vehicle can be put into an equatorial orbit by aiming it southeast from the Florida space center, then at the appropriate moment firing a rocket on board to change its orbit. That is done, for example, with communications satellites, but it is a costly, elaborate, and risky procedure. SAS-1 was to be a "bargain" mission and as free as possible from things that could go wrong. If it was launched from close to the equator, no inflight change of orbit would be needed. To have it circle the equator it would only be necessary to fire it due east.

While none of the equatorial countries had even a modest space program, a retired Italian Air Force general and engineering professor named Luigi Broglio had initiated a bold scheme. Early in the 1960s he obtained a surplus platform of the movable type used for offshore oil drilling plus an "instant pier" designed for an invasion force. Both could be floated into position, then sat on a shallow sea floor by lowering their legs. The two platforms were towed down the east coast of Africa and positioned three miles off a fishing village in Kenya a short distance south of the equator. The "instant pier," named San Marco, was used as the launch pad for rockets, such as the Scout, that could put small payloads in orbit. The drill platform, called Santa Rita, was placed several hundred meters away and linked to the launch pad by undersea cables so it could serve as the control center. For safety reasons everyone was removed from the launch facility during final stages of a countdown.

It was agreed that Italy's Centro Richerche Aerospaziale (Center for Aerospace Research), which operated the facility, would launch SAS-1 with a Scout provided by NASA. Engineers of Ling-Temco-Vought, prime contractors for the rocket, would provide technical advice.

The X-ray detectors on SAS-1, a spacecraft only one meter long, looked out opposite sides of the vehicle. One had a broad field of view so that, during the satellite's spin, it could watch any one spot in the sky long enough to pick up weak emissions. The other, viewing a smaller field, would be useful in pinpointing sources. Star sensors aligned with each detector would indicate its aim. The vehicle's circuit of the globe, at a height of 550 kilometers, would take 96 minutes.

To provide electric power, four panels extending from the craft like the vanes of a windmill would convert sunlight to electricity. The spin would be kept very slow—typically one revolution every twelve minutes—giving the detectors plenty of time to scan each part of the sky. To control the spin rate a rotor inside the craft would be kept spinning very fast—typically twenty thousand times a minute—in a direction opposite to that of the craft's own spin. If, on radio command from the ground, the rotor's speed were increased, the rotation of the craft in the opposite direction would also speed up, but because the vehicle was much more massive than the rotor, the change in revolutions per minute would be far less.

While the area of sky being scanned would depend on the aim of the spin axis, the latter could be changed gradually by activating magnets on board. These would interact with the earth's magnetic field in space to provide a gentle shove in the right direction. Since the orientation of the earth's field would be different in various parts of the orbit, it would be essential to calculate where a certain magnet should be activated to impart a proper twist to the spin axis. The X-ray observations and other readings would be stored on magnetic tape to be dumped, via radio, each time the vehicle passed over a tracking station at Quito, Ecuador.

Arrival of the Americans at the San Marco facility created a sensation on several counts. To the Italians, Giacconi said, the fact that the project manager was a woman was not only astonishing but "unbelievably exciting." That he, as scientific leader, was Italian-born was further cause for delight. The Americans found that Broglio, a man of impressive build, as befit a former general, ran the installation rather like a military post. The crew wore natty khaki shorts and knee-length white socks—in contrast to the sawed-off, tattered bluejeans and generally disreputable appearance of the

American visitors. Everyone lived ashore, at a camp near the fishing village, ferried in and out by a variety of small boats. At lunchtime they gathered on a terrazzo deck atop the control platform to feast on risotto or pasta washed down with Italian wine.

Although the success rate for launches from the platform had been excellent, the twelve-hour countdown for this one was marked by one crisis after another. The plan was to work through the night for a launch at dawn. In this way the sensitive instruments, waiting to be boosted into the cold vacuum of space, would be spared the savage heat of the tropical sun. The initial part of the countdown involved a step-by-step checkout of the many preprogramed steps that would control ignition, guidance, and detachment of each of the launch vehicle's four stages as it soared into orbit. For this, the seventy-foot, thirty-five-thousand-pound rocket lay horizontally on a massive girder that would then lift it upright. When the checkout procedure reached the point where the third stage would cease firing and back free to the final stage, the automated control system did not give the proper command.

The only thing to do was repeat the sequence in an effort to see what was amiss. This time, however, the checkout ran through perfectly. So did another—and another—and another. The NASA engineers suspected that the Italian pushing buttons and throwing switches for the checkout had, in the first test, accidentally hit a "reset" button, breaking the sequence, but the man himself was sure he had not done so. The engineers from Ling-Temco-Vought were very worried. Their contract provided that if the rocket failed, they would not receive their incentive bonus. Only if two senior NASA officials certified the vehicle as flight-worthy would they agree to the launch. Asleep in a hotel at Malindi, twenty-five kilometers away along a dirt track from the coastal village where the Italians had their base camp, were Paul Goozh, project manager for the Scout rocket program at NASA headquarters, and Roland D. ("Bud") English, who headed the Scout project at NASA's Langley Research Center in Virginia.

There was no direct communication with Malindi, but there was a radio link to the city of Mombasa, which had a telephone connection with the hotel. A guard at the hotel was

Uhuru soars into space from the San Marco platform off the coast of Kenya. (*American Science and Engineering*)

aroused and asked to awaken the NASA officials, who raced by Jeep to the village. There a motorboat awaited them, but it was planned to transfer them to a much faster rubber boat that skimmed over the waves, propelled by an outboard motor. As so often seems to happen in such a crisis, the motor refused to start.

Meanwhile, time was running out for the launch crew. They had fueled and erected the rocket in the hope that approval for a launch would be given. Every added hour of delay was a further gamble. At the outset the guidance system had been fueled with hydrogen peroxide which, in 3 per cent solution, is used as an antiseptic, but which in this case was an almost explosive 90 per cent solution used as propellant for the rocket's guidance nozzles. Now, however, the hydrogen peroxide was running low. Short test bursts from the nozzles had depleted the fuel to some extent and it was also gradually boiling away. Once it was below a critical level the launch would have to be scrubbed. An added factor was the sun, which by now was steadily rising toward the zenith, threatening to overcome the cooling system designed to protect the sensitive instruments.

Efforts to start the outboard motor and speed arrival of the NASA officials were finally abandoned and they came in the lumbering motorboat. Having climbed onto the platform they certified the launch, but it was close to 2 P.M. when the rocket roared into the sky—flawlessly. The date was December 12, 1970, seventh anniversary of Kenya's independence from Britain, and Broglio suggested that, in deference to their hosts, the satellite be named *Uhuru*—Swahili for "freedom."

A torrent of data flowed from *Uhuru*'s tape recorder every time it passed over Quito, which relayed the readings to the Goddard Space Flight Center in Maryland. From there enough "quick look" data were sent to AS & E via phone line to permit the experimenters to modify the satellite's observations if anything exciting was observed. Indeed it was. When recordings from the first seventy days of nighttime observation were analyzed, 125 sources were identified, even though only a fraction of the sky had been scanned. A few of the sources, such as Cyg X-1, had already been observed, but it was clear that the sky is, in fact, peppered with "X-ray stars." Most were so close to the Milky Way that, as previ-

Rocket launches from the San Marco platform (left) were controlled from the platform at right. (*American Science and Engineering*)

Uhuru in orbit. (*Diagram by American Science and Engineering*)

ously suspected, they were probably objects within this galaxy, but others were clearly associated with distant, eruptive galaxies. Now that more leisurely scanning was possible, many of the sources were found to be flickering in what seemed wild ways.

Six weeks after the launch the tape recorder failed and so, to obtain data from as much of the orbit as possible, the satellite was commanded to pour out its observations continuously. Stations at Singapore, Ascension Island in the mid-Atlantic and, for a time, Capetown, South Africa, recorded as much of them as possible. As analysis of the *Uhuru* data progressed, exploration of the X-ray sky became one of the scientific detective stories where a combination of clues finally falls into place, creating a comprehensive picture of events so distant one would have thought them utterly beyond the reach of human comprehension.

10
The
"Demon Bird of Satan"

Early in the *Uhuru* mission the special behavior of an X-ray source in the southern constellation Centaurus, known as Cen X-3, drew the attention of the experimenters. It had first been detected in 1967 with a rocket fired from the Hawaiian island of Kauai by a group from the Lawrence Radiation Laboratory in California. Two years later scientists from Britain's University of Leicester observed it with a Skylark rocket launched from the Australian range at Woomera. Both shots provided only fleeting glimpses, but from data collected by *Uhuru* during 74 sightings on January 11 and 12, 1971 (a month after its launching), it appeared that Cen X-3 was varying in a most peculiar manner. Using the fast-spinning rotor inside the spacecraft the American Science and Engineering group slowed the vehicle to a single revolution every 84 minutes for a new series of prolonged scans on April 10 and 12. This permitted two exposures of 150 seconds each with the wide-angle detector and six scans of 15 seconds with the narrow-angle one.

"We find," Giacconi and his colleagues reported, "that the intensity of the source indeed varies very considerably on time scales of days, minutes, and seconds." During a 150-second exposure, they said, "several peaks and valleys occur with great regularity, immediately suggesting a periodic phenomenon." A repetition rate of 4.8 seconds was sufficiently regular so that the pulses reappeared at expected times from one orbital observation to the next. Nevertheless they did vary. "The most remarkable feature of the data," the scientists reported, were "the very large changes in period occurring in short times on January 11 and 12." In less than an hour, they said, the pulse rate decreased by about a thousandth of a second, then later increased by twenty-three hundredths of a second with equal rapidity. "Orbital motions of the source about a star could not explain the abrupt period variations," they added, although they thought longer-term changes might arise from such orbital flight.

Three rapidly pulsing X-ray sources were now known: this one (Cen X-3, the Crab pulsar, and Cyg X-1. The Crab pulsar was well established as a neutron star spinning 30 times a second. The 4.8 second rate of X-ray pulses from Cen X-3 was slower than that of any known radio pulsar. The emissions from Cyg X-1 were so variable that, the experimenters said, it "may require yet a different explanation." (Indeed it did, for it was to emerge as the most likely black-hole candidate.)

In May 1971, the AS & E group observed Cen X-3 on every orbit for an entire week and was able to define its characteristics more clearly. They found that the pulse rate varied in a smooth manner—not abruptly, as previously supposed—but that the X-ray brightness of the object did change suddenly. Furthermore, both elements—the pulse rate and brightness—followed a two-day cycle (specifically, 2.087 days).

The pattern was as follows: For twelve hours the source was uniformly dim; then, for the remaining day and a half, it was bright. Its pulse rate reached its minimum during the period when it was becoming dim. It reached its maximum when it was becoming bright again.

To astronomers familiar with systems in which two stars are circling one another so closely that they are alternately eclipsed—"eclipsing binaries"—the explanation seemed fairly obvious. The X-ray source was a pulsar with a heartbeat (that is, spin rate) of 4.8 seconds circling another object, perhaps a normal star, every two days. The smooth variation of its pulse was a Doppler effect caused by orbital motion, slowing the observed rate as the X-ray pulsar moved away and increasing it when motion toward the earth reached its maximum. The twelve-hour period when the X-ray emissions dropped to one tenth their previous intensity occurred when the pulsar was eclipsed by its companion. The fact that the X-rays did not vanish altogether was taken to mean that they came in part from somewhere else, such as a gaseous envelope enclosing the two-body system and heated by the pulsar's own emissions.

"We have adopted the working hypothesis," said the AS & E scientists, "that we are dealing with an eclipsing binary system consisting of a compact object and a large massive companion. We find that all our observational data are consistent with this interpretation." The diameter of the orbit could be calculated from the lag in the pulses when the compact object was most distant. The variation in rate was so rapid, uniform, and symmetrical that a close-in, circular orbit was indicated. That the pulsar was eclipsed one quarter of the time was also evidence for its very small size. Yet its mass was estimated equal to that of the sun, whereas the other star was thought to be huge—from seventeen to forty-six solar masses.

Giacconi and his colleagues drew attention to speculation by Shklovsky and other Soviet theorists—notably Yakov Borisovich Zeldovich—on how the astonishingly powerful X rays from Sco X-1, the most powerful of all distant sources, might be generated. Shklovsky, as noted in Chapters 6 and 7, had figured prominently in the debate on quasars. Zeldovich led a group of theorists at the Institute of Applied Mathematics of the Soviet Academy of Sciences and at Moscow University. In 1943, he had been awarded a Stalin Prize—possibly for work that contributed to the Soviet weapons program. In 1939, long before the first nuclear explosion, he had done an analysis of chain reactions in uranium.

Shklovsky's explanation for Sco X-1 had been that it is a two-body system consisting of a normal star and a neutron star circling one another at close range. The superpowerful gravity of the neutron star, he said, is drawing gas from its companion, which then falls onto the neutron star at high velocity. Binary systems in which two normal stars are orbiting so closely that one robs gas from the other were well known to astronomers. It was also known that X rays, being near the shortest-wavelength (and most energetic) end of the spectrum, manifest the presence either of extremely high-energy particles whirling in magnetic captivity (as in the synchrotron radiation from the Crab Nebula first explained by Shklovsky)

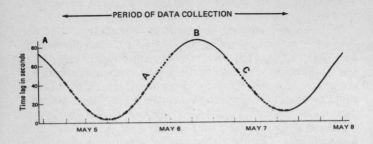

Evidence for the orbital motion of Cen X-3 is shown by the variations in arrival times of its pulses. If the source were in uniform motion relative to the earth, the rhythm of its 4.822-second pulses would appear constant. Instead their arrivals are increasingly delayed as the source moves toward the far side of its companion star (sector A on the diagram). For twelve hours the pulses are cut off (eclipsed) by the companion (sector B). Otherwise this would be the period of maximum delay in received signals. Then, as the source begins circling around to this side of the star, its pulses arrive earlier and earlier (sector C). The delay times, derived from *Uhuru* data during three days in May 1971, lie along a curve representing a 2.08-day orbit whose diameter is sufficient to produce a maximum time lag of about seventy-five seconds. (H. D. *Tananbaum* in X- and Gamma-Ray Astronomy)

or of very high temperatures. Gas falling onto a neutron star would reach a third of the velocity of light and, at the same time, would be greatly squeezed by superpowerful gravity near the star and therefore reach extremely high temperatures.

Zeldovich and his coworkers (I. E. Novikov and N. I. Shakura) pointed out that such an infall would release far more energy than if the same amount of material were used as fuel in an atomic explosion. Even the modest gravity of a body like the earth or the moon can release considerable energy from a falling body. The fall of a large meteorite, like that which formed Meteor Crater in Arizona, vaporizes the impacting object causing a great explosion. The fall of an asteroid is even more awesome, as evidenced by the concentric

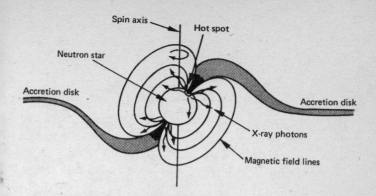

Spin axis
Hot spot
Neutron star
Accretion disk
Accretion disk
X-ray photons
Magnetic field lines

Gas falling on a neutron star whose magnetic field is tilted with respect to its spin axis generates highly rhythmic pulses, timed to its spin. Surrounding the star is an accretion disk of gas that has been robbed from the companion star (only the inner part of the disk is shown in this cross-section view). Gas from this disk is diverted toward the magnetic poles of the star, becoming greatly compressed and heated (the darker area) until it glows in in X rays near the pole. Only one pole can be seen at a time. Therefore, as the star spins, double pulses are observed. (*Adapted from* The New Scientist)

rings around Mare Orientale on the moon—a circular lunar sea surrounded by ridges and valleys in a succession of rings formed, like frozen ripples, by the explosion that also excavated the "sea." Thus, as the mass of the infalling body becomes greater, so does the impact energy. But if the force of gravity increases sufficiently, a single atom can impact with enormous energy. Even if a neutron star did not have a close companion from which it was capturing material, the Russians said, it could sweep up gas from clouds in between the stars—or even from that which had been thrown out by the supernova that formed the neutron star in the first place.

Thus efforts to explain the energy of X-ray stars—as well as quasars and explosive events far out in the cosmos— brought home the fact that, while in laboratories nuclear energy manifests the strongest force in nature and gravity is the weakest, the tables are turned where superdense concentrations of mass are concerned. Nuclear reactions in an atomic or hydrogen bomb convert less than 1 per cent of the fuel into

THE CEN X-3 BINARY SYSTEM

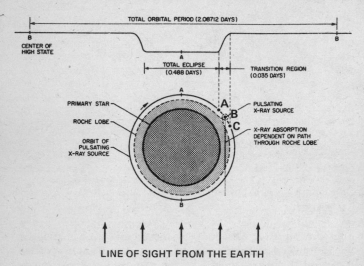

LINE OF SIGHT FROM THE EARTH

The Cen X-3 system, as deduced from its pulse behavior. The collapsed object (a neutron star) is orbiting extremely close to its companion star. The latter is surrounded by gas that is being drawn toward the neutron star, forming a so-called Roche lobe. This lobe generates a weak but continuous X-ray glow. At position A the pulsating neutron star emerges from eclipse. At position B its pulses, when beamed toward the earth, are partially absorbed by passage through the Roche lobe. At position C there is no longer such absorption and it comes into full view. The line across the top represents the variation in intensity of the observed X-ray pulses. This schematic representation is based on one prepared by the *Uhuru* group. (H. D. *Tananbaum in* X- and Gamma-Ray Astronomy) energy. An impact on a neutron star, the Russians pointed out, can release ten to twenty times that much (some theorists say even thirty times).

"Neutron stars were born at the tip of the theoretician's pen over thirty years ago," Zeldovich and Shakura wrote in 1968. Now, they said, the existence of such objects might at last be confirmed by observing the X-ray spectrum that typically would be generated by the heat of infalling gas. (As noted in the previous chapter, when X-ray sources were discovered in the early 1960s, Friedman of the Naval Research Laboratory thought they might be neutron stars. It was, however, the subsequent observation of pulsars, that most dramatically confirmed their existence.

But why should the X-ray emissions be pulsed? The explanation could be very similar to that for the radio pulsars. A neutron star should have an extremely powerful magnetic field and if, instead of being aligned with the spin axis, it was tilted (like the magnetic field of the earth), then the magnetic poles would rotate with the spin of the star. Gas falling onto the star would be drawn magnetically to those poles, generating at each of them an extremely hot spot, perhaps no more than a kilometer in diameter, that would beam X rays into space, sweeping the sky with each revolution like the lamp of a lighthouse. These emissions would be seen from a distance as pulses. If beamed X rays (or radio waves) from opposite poles were both observed, there would be two pulses, probably differing from one another in strength or other characteristics and occurring alternately—an effect clearly seen in the twin pulses of the Crab pulsar.

It appears that those pulsars throbbing exclusively at X-ray wavelengths are always in close binary systems as opposed (with one exception) to those pulsing primarily in radio waves, some two hundred of whom are now known. In a binary system the magnetic poles of a neutron star glow X-ray hot (rather than merely "white hot") because material is falling on them. A typical radio pulsar, on the other hand, would be the remnant of a supernova that has no close companion and (with rare exceptions, such as the Crab pulsar) is too run-down to generate X rays. Its emissions would be confined to radio waves because their manner of production is different, involving the ejection of particles from regions linked to the

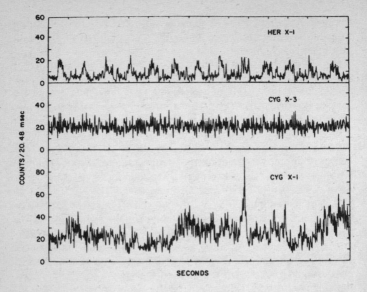

COUNTS / 20.48 msec

X-ray emissions recorded from three sources by rocket-borne instruments of NASA's Goddard Space Flight Center: Her X-1 clearly shows a pulse rate of 1.24 seconds. Cyg X-3, although notorious for wild outbursts of radio energy, has a relatively steady X-ray output, whereas Cyg X-1 (possibly a black hole) displays irregular, extremely sharp pulses.

magnetic poles. If radio pulsars lose angular momentum and slow down because of this shedding of mass, X-ray pulsars should speed up from the infalling of such material—an effect now detected in at least eight such objects.

To nail down the proposed explanation for Cen X-3, astronomers sought to find its presumed giant companion. At first they were unsuccessful, that region of the heavens being dusty, but finally, in the summer of 1973, Vojtek Krzeminski, a Polish astronomer observing with one of the new telescopes in the Chilean Andes, identified a faint star in that location whose brightness varied in a cycle of 2.087 days— the orbital period deduced from the behavior of Cen X-3.

Meanwhile, another X-ray pulsar that seemed much like Cen X-3, although weaker, had been found in the constellation Hercules. Designated Her X-1, its pulse rate was 1.24 seconds and Doppler variations in the pulsation period followed a cycle of forty-one hours (1.7 days). As with Cen X-3, there was a period during each cycle (in this case lasting six hours) when the source became very dim, as though being at least partially eclipsed. The variation in Doppler reached zero exactly at the midpoint of the dim period when the pulsar would be directly behind its companion and its motion, relative to the earth, would be entirely sideways. It again dropped to zero at the midpoint of the bright phase when it was on this side and its motion was once more lateral. The effect can be likened to that altering the pitch of a racing-car horn as observed from the grandstand facing a circular track. As the car comes around the left side of the track, approaching, its pitch reaches its highest level. As it recedes around the right side of the track its pitch drops to a minimum. But when it is passing directly in front of the stands, or opposite, on the far side, its pitch is normal, since there is no motion toward or away from the onlookers.

The novelty of Her X-1, which P. E. Boynton of the University of Washington has called "the Clockwork Wonder," proved to be an added cycle of 35.7 days, for most of which it was "turned off"—being very faint or unobservable. For 8.9 days of this cycle, the AS & E scientists reported, "the source is intense, pulsing, and exhibits the 1.7002-day intensity period." The rest of the time it "remains weak or off." To verify that this was, in fact, a recurring pattern, prolonged observation was necessary, and for five months Her X-1 kept almost continuously in *Uhuru's* sweeping field of view. The occasional interruptions included times when, because of *Uhuru's* orbit, the earth was in the way. For nine days per month from December 1971 through March 1972, Her X-1 came back to life. Several explanations for this on-off behavior were proposed. If the object is a spinning neutron star with its magnetic poles displaced from the spin axis, its pulses would be beamed to earth only as long as one of those poles, once each rotation, pointed toward us. But, as noted by Kenneth Brecher of MIT, if the neutron star were wobbling in the way tops and other spinning objects typically do, the pole might be aimed at us only periodically—in this case for nine days a month.

Another proposal was that the infall of gas from the companion star is not continuous but occurs periodically. When two stars are circling one another at close range they are not spherical; the gravity of each distorts the shape of the other so that they bulge toward one another. This bulge, known as a Roche lobe, is in effect a gravitational sack. Gas may escape from it continuously and fall onto the neutron star as with Cen X-3, whereas with Her X-1 the sack may overflow nine days a month.

The experimenters thought that seeing the companion star of Her X-1, should be easier than in the case of Cen X-3, as it was well to one side of the Milky Way's dusty clouds, but unfortunately, in November 1971—the month when Her X-1 was discovered—*Uhuru's* star sensors failed, and there was no accurate way to determine the aim of the X-ray devices. Two neighbors of AS & E—the Massachusetts Institute of Technology and Harvard University—now entered the picture. Among nine experiments on board the newly launched Orbiting Solar Observatory 7 (OSO-7) were X-ray detectors devised by George W. Clark and his colleagues at MIT, and twice, when Her X-1 was in the bright phase of its thirty-six day period, they were able to pick up its pulsations and narrow the position to within 0.3 degree.

While a great many candidates would lie within so large a circle, William Liller of the Harvard College Observatory suggested that astronomers look at an unusually blue star known as HZ Herculis (from the Humason-Zwicky catalogue) which, as early as 1936, had been observed to be variable in an irregular and rapid manner. Within a few weeks two young astronomers at the University of Rochester, Donald Q. Lamb and John M. Sorvari, decided to see if anything in the vicinity of the star was flashing optically at the 1.24-second rate observed by *Uhuru* in X rays. They used equipment that had been developed shortly after the discovery of pulsars in the hope of detecting optical flashes from such objects. The timer on the system was set to match the X-ray pulse rate and on the night of July 8, 1972, three separate observations were made from the university's Mees Observatory atop Gannett Hill near Naples, New York. Fluctuations in light brightness from the star or its neighborhood did, indeed, seem to follow the X-ray tempo, but when the location was scanned three and four nights later, no pulses were observed.

The two men reported their observations in the urgent manner reserved for such discoveries—via the internationally sponsored Central Bureau for Astronomical Telegrams in Cambridge, Massachusetts. A number of observatories attempted to confirm the report, but none were able to do so. The Rochester astronomers stuck to their guns, nevertheless. They felt sure the pulses they had seen were real. Finally, many months later, a group from the University of California at Berkeley found that HZ Herculis is, indeed, pulsing. However, its optical pulses differ from those of the X-ray pulsar, apparently because they originate in pulsed illumination of the star by X rays from the orbiting pulsar. The result is complicated modulation of the optical pulse rate. Earlier attempts at verification had failed simply because—as in initial attempts to detect the Crab pulsar—the timer was not adjusted to allow for these effects.

Almost simultaneously with the first detection of optical pulses, John N. Bahcall and his wife, Neta, began a series of eighty-one brightness measurements of HZ Herculis using the one-meter reflector at Tel Aviv's Wise Observatory in the Negev desert. Each was a fifteen-minute observation made on clear nights from July 6 to August 19. They found that the star's brightness varied in a cycle identical to the forty-one-hour variation in X-ray pulse rate. But the star obviously was not being eclipsed. It turned out that HZ Herculis was dimmest, not when the pulsar was in front of it, but when on the far side. The two objects, therefore, were both brightest at the same time in the cycle.

When they thought about it, the Bahcalls realized this was not surprising. The pulsar, if a neutron star, would be so small that its passage in front of the companion star would cut off a negligible amount of its light. But if the pulsar were extremely brilliant in X rays, those rays would heat the companion star, making it shine (and pulse) brightly on the side facing the pulsar.

Meanwhile Liller and his colleagues, observing HZ Herculis through Harvard's 1.55-meter reflector near the village of Harvard, Massachusetts, looked for evidence in its light of the monthly cycle of on-off behavior seen in the X-ray pulsar. Neither they nor the Bahcalls found any such variation, implying that illumination of the star by its X-ray companion continues unabated. This strengthened the suspicion of the Harvard group that the long periods when the pulsar faded arose from wobbling, or precession, of its spin axis. "If the magnetic axis were tilted with respect to the spin axis," they wrote in a letter to the *Astrophysical Journal,* "then one would expect a pulsed signal to be emitted from the [neutron] star's surface (or near it) with a pulsation period equal to or half that of the rotation period of the neutron star." (The pulse period would be half the star's rotation period if pulses from both poles were observed.)

At times, they said, "the Earth may lie completely out of the cone of X radiation emitted by the hot spot." The effect of all these factors, they noted, was to make HZ Herculis seem a most peculiar star, alternately displaying surface temperatures of two completely different stellar types (when cool an F4 subgiant and, when hot, a B8 main-sequence star). Its cool side was presumably a better index of its proper classification.

Liller and his colleagues went to the old Harvard plates and found that over the years since 1890 HZ Herculis had gone through periods of months or years when it remained continuously dim. At the conclusion of their letter they wrote: "We urge optical astronomers to make every effort to undertake detailed studies of this most incredible system."

Boynton, at the University of Washington, suggested that Nature might be baffling us with a surfeit of information on the system. "Perhaps," he said, "we are privileged to see too much in the sense that we have to face the additional task of separating trivia from the essential. Is the Hercules system a gaudy freak immodestly revealing extraneous complexity, a misinterpreted Rosetta Stone of mindless doubts?" Or does it offer the best insights into such systems? He left the question open.

By 1979, with much larger and more sophisticated spacecraft in orbit, roughly a dozen X-ray pulsars had been discovered and in a number of cases their locations pinned down sufficiently to identify their visible companions. Early in 1978, the first High Energy Astronomy Observatory

(HEAO) was swung from its designed orientation so that for two ninety-minute orbits it could watch a newly found, 3.6-second pulsar in the constellation Cassiopeia. Normally HEAO spins around an axis that keeps its power-generating solar panels facing the sun, but this somewhat risky maneuver, under radio command from the ground, made it possible to narrow down location of the pulsar so that its optical counterpart could be identified.

Among the most intensively studied such pulsars have been Sco X-1, the most powerful of all, and Cyg X-3, notorious for its wild outbursts at radio frequencies. Within two days in 1972, the radio brightness of Cyg X-3 increased one thousand fold. It seems to be wobbling in a seventeen-day cycle comparable to the one of 35.7 days manifested by Her X-1. Some pulsars have been found with rates as slow as several hundred seconds, which is puzzling since neutron stars should spin far faster. They may be less dense than neutron stars or have been slowed by interactions between their magnetic fields and clouds of surrounding electrified gas.

Although Sco X-1 is the most powerful X-ray source, efforts to find it or a stellar companion with optical telescopes were frustrating. A number of astronomers tried to detect rhythmic behavior indicative of orbital motion in a flickering, bluish star found at its location in 1966. A Soviet group, after monitoring its brightness for two hundred nights, thought it saw a cycle of just under four days. A half-day cycle was reported by a Dutch astronomer, and two Canadians, from spectral observations, deduced a period of only six and a half hours.

Liller and colleagues at the Harvard-Smithsonian Center for Astrophysics decided the only way to avoid deception by short-term variations was, with a computer, to digest "a large amount of data covering a long period of time." They analyzed the star's brightness, relative to its neighbors, in 1,608 photographs of the region made by Harvard observatories from 1889 to 1974. Using additional data from other sources they fed 1,766 brightness measurements into a computer, which tested the data to see if they fit any cycle from a quarter day to one hundred days. The result, the Harvard-Smithsonian group reported, pointed clearly to a period of 18.9 hours. Sco X-1 is thought to be a small collapsed object (white dwarf or neutron star) circling a small normal star at such close range that large amounts of gas are being drawn from the normal star. It is this envelope of hot gas that is observed optically, rather than the star itself. The large volume of gas that falls on the collapsed object accounts for the intense generation of X rays. The latter show no evidence of a pulse period.

A footnote to the observations of Sco X-1 has been the discovery that its emissions are so strong they apparently affect long-range communications on earth. In 1960 the Physical Research Laboratory in Ahmadabad, India, began recording a powerful Soviet radio station that broadcast from Tashkent a steady transmission at 164 kilohertz so that distant observers could record changes in radio-reflecting properties of various layers of the upper atmosphere, such as those caused by eruptions on the sun. Between April and July, from 1960 to 1963, the Indians noticed a sharp drop in signal strength that occurred progressively later each night. Tashkent lies some 2,000 kilometers north of Ahmadabad and it was found that the signal drop occurred each time Sco X-1 passed over the midpoint where the Tashkent signals would normally be bent down toward Ahmadabad. As reported in *Nature* (by S. Ananthakrishnan and K. R. Ramanathan), X rays from Sco X-1, as that source passes overhead, apparently ionize the D layer of the atmosphere 75 kilometers aloft sufficiently to prevent radio waves from penetrating to the higher layer that normally bends them back toward the earth.

The most exciting developments, climaxing exploration of the X-ray sky, have been related to Cyg X-1, whose emissions (at times second in strength only to Sco X-1) were first detected by the 1962 Aerobee flight that many regard as having given birth to X-ray astronomy. Observing Cyg X-1 originally produced considerable anguish among astronomers because each group obtained—and published—contradictory reports on its behavior. The Naval Research Laboratory, with its rockets, gained two brief glimpses of the source in June 1964 and April 1965. On the second occasion its brightness had dropped 75 per cent. Cygnus X-1, the Navy scientists reported, "is the first clear example of an X-ray variable. It cannot be specified how rapidly the variation occurred, only that it occurred between the observations. . . ."

During *Uhuru's* early sweeps of the sky it viewed the Cygnus region for twenty seconds at a time. Pulsations were observed that, while varying, seemed to fit an underlying rate of .073 second. But other observers disagreed. A group from NASA's Goddard Space Flight Center reported from eight seconds of rocket observation in 1970 evidence for two pulse rates: at 0.290 second and 1.1 seconds. An MIT team, with seventy-five seconds of data from a 1971 rocket flight, could find no such pulses, although there were flareups as brief as .05 second. This was essentially confirmed by those at the Naval Research Laboratory who reanalyzed forty seconds of data from a 1967 flight.

It was decided to take a long look with *Uhuru,* and one thousand seconds of data were recorded during the six months ending in June 1971. "The characteristic behavior of Cyg X-1, as revealed by this study," said the AS & E group, "is consistent with all of the above observations." In other words, everyone was right. Summarizing, they said:

"1. Large fluctuations of intensity exist on all observed time scales ranging from fifty milliseconds [thousandths of a second] to ten seconds containing up to 50 per cent of the power.

"2. Periodic pulse trains with periods [repetition rates] from 0.3 second to over ten seconds exist containing 10–25 per cent of the power; however, they persist only for several seconds or several tens of seconds.

"3. No single period is consistently present."

Thus the pulses come in short trains, each with its own tempo, and the rates neither last very long nor recur. Similar behavior, the report said, was being recorded from Cir X-1 in Circinus, the constellation of the Dividers or Drawing Compasses. "The data show large pulsations," it added, "with no obvious stable period. . . .No generally accepted theoretical model yet exists to explain the behavior of this class of objects."

Perhaps, the radio astronomers thought, they could learn

something about them. The most powerful of all X-ray sources, Sco X-1, was observable with radio antennas as a strong, highly variable source and for a year at the National Radio Astronomy Observatory Robert M. Hjellming and Campbell M. Wade sought to pick up emissions from a half-dozen other X-ray sources, including Cyg X-1. They used the observatory's antennas at Green Bank in a lonely West Virginia valley well protected from radio interference. When they looked toward Cyg X-1 (its location known only approximately from *Uhuru* data) on June 14, 1970, and March 21 and 22, 1971, nothing was recorded, but beginning May 13 of that year they repeatedly observed emissions.

Meanwhile, in the Netherlands, L. L. E. Braes of Belgium and G. K. Miley, an Irishman, were conducting a similar search, using the long row of dish antennas at Westerbork that (with the interferometry method devised by Sir Martin Ryle at Cambridge) provided a radio map of the area under observation. On February 28, 1971, they detected nothing, but when they next looked, on the night of April 28–29, Cyg X-1 had burst into view.

Thus some time between March 22 (when nothing could be observed from Green Bank) and April 28 (when it was detected by Westerbork) Cyg X-1 had "turned on" for the radio astronomers. Yet at the same time, as recorded by *Uhuru*, its X-ray emissions decreased to one quarter their previous intensity, although remaining highly variable. This continued four years until in the spring of 1975 scanning by the British satellite Ariel 5 and by the Astronomical Netherlands Satellite indicated that the intensity at longer X-ray wavelengths briefly increased. A few days earlier Aryabhata, India's first earth satellite, had been launched (with Soviet help) and obtained a last-minute look at the X-ray emissions before the transition. While the shift to a state in which it was brilliant in X-rays and dim at radio wavelengths was short-lived, a more prolonged changeover occurred six months later, ending in February 1976, when it again became dim in X rays and bright in radio waves. This transition, according to Braes and Miley in the Netherlands, was "remarkably similar" to that seen in 1971. Here, then, was another puzzle to be explained by the theorists—or clue to help them guess the nature of Cyg X-1.

Once the Green Bank and Westerbork radio astronomers

Using the method developed by Sir Martin Ryle and known as "aperture synthesis," the antennas of the Westerbork Observatory in the Netherlands are able to map regions of the sky at radio wavelengths. A form of multi-antenna interferometry with computer assistance is used. The antennas helped pinpoint the location of Cyg X-1, the leading black-hole candidate. *(Aerofoto Eelde)*

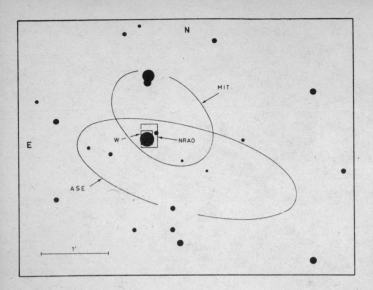

Pinpointing the location of Cyg X-1 and showing its relationship to the giant star HDE 226868 was achieved in a series of steps narrowing its location in the sky. The large oval shows the region indicated by the satellite *Uhuru*. The more round oval was the MIT rocket determination, also at X-ray wavelengths. The larger box was determined by the National Radio Astronomy Observatory and still included a small, somewhat variable star as well as HDE 226868. The smallest box, obtained by the Westerbork antennas, included only the giant star. (*From a paper by L. L. E. Braes and G. K. Miley in* X- and Gamma-Ray Astronomy).

had picked up Cyg X-1 in 1971 and narrowed down its location, it was inevitable that optical astronomers should look to see if anything there was shining in visible light, and again the Big Bertha of optical astronomy, the five-meter reflector on Mount Palomar, was called into service. An observing team that included Jerome Kristian and Allan Sandage had five independently derived positions to guide them: two from the radio astronomers and three—less precise—from X-ray observations by *Uhuru* and by rockets carrying instruments from MIT and the Lawrence Radiation Laboratory (one had also been obtained by a University of Tokyo balloon). When the uncertainties were taken into account, each "position" was actually an "error box" within which the object was thought to lie. The error boxes provided by the five groups overlapped and in the area included within all but one of them were two candidate stars. One was listed as HDE 226868 in the *Henry Draper Extension,* a sequel to the great *Henry Draper Catalogue* of some 360,000 stars, most of them classified according to their spectra by Annie J. Cannon, a devoted member of the Harvard College Observatory staff. The project, initiated in the closing decades of the last century, had been made possible by a fund memorializing Henry Draper, a distinguished physician and amateur astronomer.

HDE 226868 was classed a normal B-type star. If so, it would be very large and intrinsically very bright. Since its observed brightness was not great, this was taken to mean it was distant—more than 6,500 light-years away. Yet, from the amount of low-energy X-rays that were getting through the interstellar hydrogen assumed to lie between *Uhuru* and Cyg X-1, Herbert Gursky and others at AS & E had estimated it was no more than half that distance. The Palomar Observatory therefore focused its attention on the other candidate star, which seemed unusual and variable.

The radio astronomers, however, were still at work and both Green Bank and Westerbork soon had better positions. It became clear that the emissions were from the direction of the "normal" B-type star. "The nearly perfect positional coincidence of the radio source and the star," Wade and Hjellming reported in *Nature,* "leaves no doubt that HDE 226868 is the correct identification."

Meanwhile, two observatories had turned optical telescopes on that star to see if its spectrum was varying the way it would if it were circling a companion. At the Royal Greenwich Observatory (which after World War II had moved from Greenwich to Herstmonceaux Castle in Sussex to escape the lights of London) Louise Webster and Paul Murdin, using the 2.5-meter Isaac Newton Telescope, found evidence that HDE 226868 was, indeed, circling something every 5.60 days. They concluded that the visible star was somewhere between ten and thirty times more massive than the sun and that the invisible companion was probably between 2.5 and 6.0 solar masses. The companion might be a neutron star or white dwarf, they said, but since it seems more than twice as massive as the sun—well in excess of theoretical limits for white dwarfs or neutron stars—"it is inevitable that we should also speculate that it might be a black hole."

Parallel observations were being made in Canada by C. T. (Tom) Bolton of the University of Toronto with the 1.88-meter reflector of that university's David Dunlap Observatory at Richmond Hill, Ontario. Between mid-September and mid-November 1971, he obtained a number of spectra, and others were taken for him by the Steward Observatory in Arizona. The combined results indicated orbital motion virtually identical to that deduced by the Royal Greenwich Observatory.

Furthermore, he said, there was spectral evidence that gas was streaming from the visible star toward its companion. The orbital movements of HDE 226868 indicated that it was locked in gravitational embrace with another object of considerable mass. As a B star, Bolton argued, it could not reasonably be less than twelve solar masses, and the lower limit on the invisible companion or "secondary" must therefore be three solar masses. His findings thus were very close to those of Webster and Murdin. His report to *Nature* said, in part:

> The motion of the gas stream and the apparent position of the X-rays are being produced through the interaction of the gas stream and the unseen companion. The high energies of the X rays imply that large accelerations are involved—accelerations such as might be produced by the intense gravitational field of a collapsed object. The lower limits placed on the secondary mass are too high for a white dwarf and probably rule out a neutron star. The lack of a supernova remnant also argues against the secondary being a neutron star. This raises the distinct possibility that the secondary is a black hole.

Finally, members of the Moscow group (A. M. Cherepashchuk, V. M. Lyutin, and R. A. Sunyaev) reported that the brightness of HDE 226868 was varying in the 5.6-day cycle. As pointed out by Bolton in his summary of the accumulated evidence in the December 11, 1972, *Nature,* the variations in brightness probably occurred because the gravity of the unseen companion had pulled the star out of shape so that its size, as seen from the earth, varied during each rotation. By now Bolton was estimating the mass of the invisible object as greater than 7.4 solar masses, assuming 30 solar masses for HDE 226868. The sudden decline in the brightness of Cyg X-1 in April 1971, he added, may have occurred when the bulge of the visible star shrank and fed less gas into the stream plunging toward its companion. Analysis by others indicated that Cyg X-1 was orbiting extremely close to the star—at less than one third the distance of Mercury, the innermost planet, from the sun. No wonder it was robbing the star of gas!

That the streaming of gas from the visible star is highly erratic was indicated by a series of observations made from 1972 to 1975 by British X-ray detectors on the Copernicus astronomical satellite. When the invisible object went behind HDE 226868 its X rays were absorbed in an irregular manner, as though from passage through such variable gas streams.

The accumulating evidence on Cyg X-1 soon convinced Giacconi and his colleagues that it is a black hole. Its X-ray emissions, instead of being pulsed, as with a spinning neutron star, fluctuate wildly on time scales as short as a thousandth of a second, indicating an extremely compact source region. Its mass, as noted by Harvey D. Tananbaum of the AS & E group at a 1972 symposium of the International Astronomical Union in Madrid, "taken conservatively" must be three times that of the sun.

According to the tradition of reasoning, from the limit on white dwarfs set by Chandrasekhar to those placed by Oppenheimer on neutron stars, anything so massive must continue to collapse indefinitely, fulfilling all the theoretical predictions of infinite density, such as a pinching of space and time out of existence.

Many astrophysicists therefore agree that Cyg X-1 is a black hole, although some, to hedge their bets, say simply that this is the least contrived explanation of the observations. Others have been less inhibited. A NASA press release has reported that British scientists, based on data from the Copernicus satellite, "say they have established that the previously predicted black hole in space is no longer theoretical. It is a fact." Yet Stephen Hawking, who has probably theorized on black holes more intently than anyone, is himself not convinced. He has made a bet with Kip Thorne of Cal Tech that Cyg X-1 is not a black hole (specifically one with a mass greater than the Chandrasekhar limit). If he wins, he receives a four-year subscription to *Private Eye*. If Thorne wins, the prize will be a year's subscription to *Penthouse*. It is a bet that may not be resolved for a very long time.

One inquirer into the most ancient histories of the human race points out that the region of sky occupied by Cyg X-1 seems to have been held in special awe by the Sumerians, some five thousand years ago. George Michanowsky, who has studied clay tablets inscribed with the cuneiform records of that civilization, notes that the constellation Cygnus was called Ud-Ka-Duh-A, referring to a panther-griffon "demon bird" associated in Mesopotamian myth with Nergal, lord of the netherworld. In his book *The Once and Future Star* he says:

> This coincidence is made even more remarkable by the fact that celestial coordinates suggested by some cuneiform texts specifically place Ud-Ka-Duh-A in a sky region dominated by the Star Delta Cygni. It is precisely in that area that Cygnus X-1, the first black-hole phenomenon to be discovered, has been located.

The Sumerian inscription that, according to Michanowsky, refers to the Cyg X-1 region of the sky in terms of a "demon bird" associated with the lord of the netherworld.

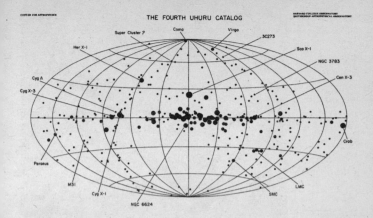

The X-ray sky as seen from *Uhuru*. This is from the fourth *Uhuru* catalogue, compiled in 1977. The sky is shown with the Milky Way corresponding to the equator in maps of the earth. The core of the galaxy is at the center. *(Center for Astrophysics)*

His implication is that the Sumerians observed an awesome occurrence in that part of the sky associated, perhaps, with the collapse of Cyg X-1 into a black hole. Whether it and similar objects such as Cir X-1 are black holes or have stopped short of total collapse, there is no doubt that they provide glimpses into extraordinary phenomena seen nowhere else. When Giacconi, as one of the first explorers of the X-ray sky, came to summarize the early discoveries he spoke of "the feeling of awe and gratitude that many of us have shared as this rich new universe of observational facts has revealed itself to us; awe at the infinite richness and variety of nature which so greatly has surpassed our wildest speculations, and gratitude at having lived in what has been one of the heroic periods of a new branch of observational astronomy."

11

The Tests: Prediction and Observation

Usually theories are developed to explain things that have already been observed, but once in a while a theory goes beyond that and predicts something that has not been seen. Its observation then provides dramatic confirmation of the theory. So was it with the discovery of neutron stars. And so is it—perhaps—with black holes.

Nevertheless, the idea of a situation where, in effect, time and space lose their identity is so wild that some theorists are reluctant to accept it. While the properties of such objects are predicted by Einstein's general theory of relativity, the skeptics suspect there may be holes in the theory rather than in the sky. They point out that scientists in seeking to explain phenomena in terms of a mathematical theory sometimes find that their formulation of it predicts situations where one or more factors become infinite. In the past this has always been a warning sign that they were on the wrong track and had to start over again. And no theory predicts so many infinities as that of the black hole.

As a result the skeptics have tried—some would say almost desperately—to explain Cyg X-1 and its sister objects without resort to black holes, one of their arguments being that some theories of gravity do not predict such an extreme end situation. At the same time those more receptive to the concept have been searching for other examples and more convincing evidence.

Beginning in 1973 a variety of explanations for Cyg X-1 that avoided recourse to a black hole were proposed. One was

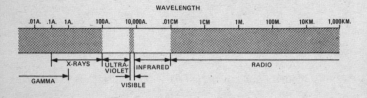

The electromagnetic spectrum extends from the longest waves, at the radio end of the spectrum, to the shortest (X rays and gamma rays). Visible light occupies only a narrow band near the center of the spectrum. While gamma rays occupy the extreme short-wave end, representing the most energetic emissions of all, there is a certain overlap with the X-ray region because, strictly speaking, X rays are defined as emissions from highly energetic electrons, colliding, for example, in a very hot gas, whereas gamma rays are products of nuclear reactions, as in radioactive decay and matter-antimatter collisions.

that the X-ray source is a neutron star orbiting a massive normal star of ten solar masses which, in turn, is in orbit around the giant HDE 226868. The evidence that the latter's companion is too massive to be a neutron star would then apply to this third member of the system (the one of about ten solar masses). A similar explanation was advanced by a group that proposed, as well, an arrangement in which the two big stars were in close orbit around one another and the neutron star (Cyg X-1) circled them both.

In another line of attack three scientists at the University of Maryland argued that identification of Cyg X-1 as a black hole was based on "a truly staggering number of assumptions," one of which was that HDE 226868 is really a supermassive star. They cited recent studies of star HZ 22, whose spectrum was quite similar but which was only about half as massive as the sun. Such stars, they said, are exceptional, and this was a "less likely" description of HDE 226868, but it could not yet be ruled out. If the star were, in fact, like HZ 22, it would be intrinsically dim and therefore relatively close (3,300 light-years) rather than brilliant and twice that distance, as had been supposed. Being less than the mass of the sun, its companion would have far too little mass to be a black hole.

This led two teams of astronomers at the Lick Observatory in California to scan that region of the heavens for clues to the star's true distance. They both concluded that it is, in fact, very far away and, as one of their reports said "models that invoke a black hole for the collapsed secondary become more tenable."

The most persistent of the dissenters were at the Massachusetts Institute of Technology, almost next door to those at American Science and Engineering, Harvard, and the Smithsonian, who were ardent proponents of black holes. The MIT scientists argued, for example, that, from observations with their instruments aboard Small Astronomical Satellite 3 (SAS-3), which, like *Uhuru*, had been launched from the platform off Kenya, the mass of the X-ray pulsar in the constellation Vela is at least 1.7 times that of the sun, whereas it had been widely assumed that anything that massive had to be a black hole. This object clearly could not be such since, in that case, it would not produce rhythmic pulses. A black hole cannot have any surface features, such as the off-center magnetic poles needed to generate pulses. The implication was that this invalidated the assumption that anything much larger than one solar mass had to be a black hole.

In rebuttal it was noted that some factors could permit neutron stars to be a little (but not much) larger than one solar mass. Remo Ruffini and Robert W. Leach, his student at Princeton, proposed that they could, at most, be 3.2 times the mass of the sun.

The MIT skeptics also raised a troublesome question about how a system in which two stars are circling one another at close range could survive the catastrophic explosion marking the collapse of a supergiant star into a black hole. Why was the companion star not blasted off at high speed? A number of fast-moving "runaway" stars are believed to have been propelled in this manner. It was remotely possible that such collapses do not always produce great explosions, said Kenneth Brecher and Philip Morrison, but, they added, "it is difficult to understand the highly circular orbits of the stars in Her X-1 and Cen X-3," though, they said, tidal forces might help make the orbits circular again. They proposed that the binary X-ray sources were collapsed stars of white-dwarf dimensions spinning at extremely high speed.

The question of how two-star systems remain closely bound, even though one star may have become a supernova, also bears on the discovery by astronomers of the Five College Observatory in Amherst, Massachusetts, of a radio pulsar (presumably a neutron star) within such a system. An early analysis of the problem of supernovas in close binaries was done in 1969 by Stirling A. Colgate, who a decade earlier had undertaken an intensive theoretical study of supernova explosions. He had been motivated to do so by his observation that those at Geneva discussing a ban on nuclear-weapons tests in space were not sure how to discriminate between such a test and a nearby supernova.

Colgate (descended from William, whose New York soap factory, established in 1806, evolved into a far-flung toiletries industry) concluded that a supernova would not necessarily blow off the companion star. Others pointed out that the evolution of a black hole in such a system could take place in two stages. The first would be collapse to a neutron star, followed by steady robbing of gas from the companion star until the mass of the neutron star exceeded the black-hole limit and it collapsed further. If these events were explosive, it was proposed, they would distort the orbits of the two stars, making them eccentric rather than circular, but tidal and magnetic drag and gas exchange between the two bodies would gradually circularize the orbits again (assuming they were close enough to one another for such interactions).

In any case, it seems clear that one member of a tight, two-star system can collapse to a neutron star without blowing the system apart. Otherwise none of the X-ray pulsars, believed to be neutron stars in close orbit with a companion, would exist. That the process may also be less violent than the more catastrophic supernovas is hinted at by the failure, so far, to observe any expanding gas bubble, such as that of the Crab Nebula, centered on one of the X-ray pulsars.

Probably the most novel alternative to black holes considered by the MIT group was inspired by theoretical developments there relating to the behavior of very closely packed nuclear particles and to the evidence that they are formed by subunits widely known as quarks. For reasons that theorists would like to explain, bombardment of nuclear particles (protons and neutrons) show that there are "things" inside of them, but no one has been able to knock those "things" loose and study them. They appear to be the hypothetical quarks, and Robert L. Jaffe, Kenneth Johnson, and their MIT colleagues have proposed that quarks are confined inside a bubblelike "bag" from which it is intrinsically impossible to extract them. The concept is known as the "MIT bag model" and it has been suggested that the collapse of a massive star would produce an enormous, incompressible bag of quarks— a "quark bag star."

At the Eighth Texas Symposium on Relativistic Astrophysics, held at Boston in 1976, Brecher and G. Caporaso, citing such ideas, said: "Neither the local properties of matter at supernuclear densities, nor the effects of gravitation when fields are large are *experimentally* known." Furthermore, they said, "Using experimentally allowed theories of matter and gravitation," the value for the maximum mass of a "neutron" star (which could, in fact, be a bag of quarks) might lie anywhere between one and eighty solar masses.

In a lighter vein the MIT theorists produced a one-page newspaper called *The Black Hole,* which looked suspiciously like *The New York Times* (which had carried a number of reports on the black-hole proposals). In addition to various reports of alleged evidence for black holes, it described the Brecher-Morrison proposal that an X-ray source like Cyg X-1 may actually be a fast-spinning dwarf star formed of collapsed ("degenerate") matter circling a normal (and therefore younger) star. The headline on this item read:

Exciting Young Star Finds Happiness
With Old Degenerate Dwarf

Below was an illustration captioned: "First detailed color photograph of a black hole. Note features at upper left and center, in good agreement with current theoretical predictions." The entire picture was an expanse of featureless black.

One way to strengthen the argument for black holes would be to find more objects best explained in this manner. The search actually started as early as 1964 when Zeldovich and Guseynov in the Soviet Union began studying catalogues listing the several hundred known "single-line binaries." These are two-star systems in which one of the pair is invisible. They are spectroscopically "single line" systems because in their light only one version of each spectral line is evident, that line shifting back and forth in wavelength as the star moves toward and away from the earth in its orbital flight. In cases where both stars are visible, two lines appear whenever one star is approaching and the other receding. In most cases, the Russians assumed, one star would be invisible simply because it was too small and dim to be seen at such a distance or because its light was overwhelmed by that of its companion. But if analysis of the orbital movements showed the unseen object to be very massive, then perhaps it would be invisible because it was a black hole. They found five such candidates.

Then at the California Institute of Technology, Kip S. Thorne and Virginia Trimble carried the search further. "Unfortunately for us," Thorne reported, "none of the eight good candidates on the new list we prepared presented a truly con-

[216]

Les Corps Obscurs de Laplace-Existent-Ils?

PARIS, 1796 — At a recent meeting of L'Academie des Science, M. Le Marquis De Laplace, the eminent mathematician and natural philosopher, provided for all those present a most amusing and entertaining evening. With readings from his recent best seller "Exposition Du Systeme Du Monde," while circulating amongst the audience

Book Review:
"The Other Side"
by Alfred Kubin

VIENNA, 1909 — In a fit of brilliant insight and intense productivity, the great Austrian presurrealist painter Alfred Kubin has succeeded, where no man has before him, in grasping the full physical significance of collapse into a black hole. A brief illustration from his novel "Die Andere Seite" should suffice to support this claim. Turning from the brush to the pen, he wrote: "And now, for the first time, I discovered in the veil of mist an immense, high wall. Suddenly, unexpectedly, it loomed up before me. Someone carrying a light was walking in front of us toward an enormous black hole: that was the gate to the Dream Kingdom. As we approached I noticed its huge dimensions. We entered a tunnel, keeping as close as we could to our guide. Then something strange happened. I had already penetrated some distance into the waulike passage when I was overcome, as though at a blow, by a wholly unfamiliar and dreadful sensation. It began at the back of my head and ran down my spine; my breath stopped, and my heart beat wildly. Helplessly I looked toward my wife, but she herself was white as a corpse, deathly fear mirrored in her face. In a quivering voice, she whispered: 'I shall never come out of here again.'" His recognition of the role of tidal forces and of the irreversibility of such a predicament are all the more remarkable for they predate Herr Einstein's General Theory of Relativity by seven years.

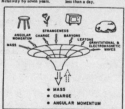

A rose is not a rose, nor would it smell as sweet, were it to be inside a black hole whose only attributes are mass, charge, and angular momentum.

reprints of his latest paper in the Allgemeine Geographische Ephemeriden, he presented a talk entitled "Future Progress of Astronomy." Amongst other speculations, he suggested that the Universe is filled with "des corps obscurs," dark bodies, in numbers equal to the visible stars! He bases these ideas on his calculations which show that "a luminous star, of the same density as the earth, and whose diameter should be 250 times larger than the sun would not, in consequence of its attraction, allow any of its rays to arrive at us." He concluded by saying that "it is therefore possible that the largest luminous bodies in the universe may, through this cause, be invisible." Despite the irrefutability of his mathematics, he failed to suggest how any object would come to exist in such an ignominious state. One can only hope that his good name will not be delivered by such flights of fantasy. (Ed. note — By the publication of the fifth edition of "The System of the World" Laplace had expunged all references to "des corps obscurs.")

SCIENTISTS FORESEE:
COLLAPSE INEVITABLE

BERKELEY, 1939 — Out of the depths of the Great Depression, and confronted with the possibility of another worldwide conflagration, the young American physicist J. Robert Oppenheimer and his graduate student, former truck driver Hartland Snyder, have reported in the latest issue of the Physical Review that "when all thermonuclear sources of energy are exhausted, a sufficiently heavy star will collapse." Such news should be kept in mind by those who would hope that a detente could be achieved by bringing pressure to bear on arbitrarily large bodies to counter the ever present gravity of the situation. Furthermore, as the authors are the first to point out, while a sufficiently distant observer will never see its final demise, a person collapsing with a massive body will experience all the accompanying stresses in less than a day.

A rose is not a rose, nor would it smell as sweet, were it to be inside a black hole whose only attributes are mass, charge, and angular momentum.

CYGNUS X-1: BLACK HOLE OR RED HERRING?

Popular model of Cyg X-1, consisting of a binary star system containing a black hole (at the center of the disk, lower left) accreting matter ejected from its more massive companion.

Exciting Young Star Finds Happiness With Old Degenerate Dwarf

BOSTON, 1973 — On a day with very little news reaching us, a hopeful and touching story has emerged. It is commonly believed that overweight old stars have no alternative but to eventually collapse and disappear from sight altogether. Not so, say two MIT Professors, K. Brecher and P. Morrison. In a surprising twist of the usual scenario, they suggest that such stars can avoid this fate by turning instead into degenerate dwarfs. If they get around enough, such stars can again become radiant and even, as they suggest in the case of Cygnus (The Swan) X-1, cohabitate with a star as young and bright as HD226868. (Ed. note — This story should satisfy those readers who have accused us of a discriminatory publishing policy. It is only the first in our new affirmative action series featuring such recently neglected stars as white dwarfs, red giants and, if space permits, blue stragglers. This series will complement our ongoing reports on the activities of some prominent white holes. Owing to cosmic censorship, however, we have been unable to uncover any information surrounding naked singularities!)

Princeton Professor Proclaims Black Holes Have No Hair

PRINCETON, 1972 — Professor John Wheeler of the Princeton University Physics Department, reporting on his own researches, as well as those of Drs. Penrose, Hawking and others of Great Britain, has revealed that should black holes be discovered soon, there is little to distinguish one from the other. This follows, he says, from very general and powerful mathematical theorems which imply that such a body is completely independent of three independent quantities: mass, charge and angular momentum (see figure). Such a conclusion, however, may be premature as has been emphasized by Professor F.

First detailed color photograph of a black hole. Note features at upper left and center, in good agreement with current theoretical predictions.

Curtis Michel in his recent article in the journal Comments on Astrophysics and Space Physics entitled "Hair For Black Holes." He cautions the unwary, "If black holes indeed have no hair, it could be because they have no scalp for it to grow out of. However, there is a lot of stuff floating around looking suspiciously like dandruff."

NEW YORK, April 1, 1971 — The New York Times today reported for the first time the discovery of a "black hole in space." Variously referred to as a "collapsar" (in G.W. Cameron of the (Yeshiva) Center for Astrophysics) or "frozen star" (Ya.B. Zeldovich of the Soviet Academy of Sciences), such objects have long filled the void of theoretical astrophysicists waking hours. Now at last, it seems, there is an object upon which they can lavish their speculations. Scientists from American Science and Engineering, Inc., headed by Dr. Riccardo Giacconi, making observations with instruments aboard the first small astronomical satellite, nicknamed UHURU, claim to have finally shed some light on the matter of black holes or, more precisely, say that they have seen the light, from matter spiralling headlong into the oblivion of a black hole. They interpret the x-ray emissions from Cyg X-1 as arising from gas flows in a close binary star system containing a massive young star ejecting luminous matter, which then accretes onto its fully collapsed companion (see picture). Waving aside the objections of a dissident minority of scientists who question whether Cyg X-1 is fully collapsed, or massive, or accreting, or even, whether it is in a binary star system, Dr. Giacconi told this reporter in no uncertain terms that "..."

(Continued on page 13)

Texas Teachers Tout Tunguska Tragedy

AUSTIN, 1973 — Waving aside as extravagant and speculative the claims by Russian scientists that the immense explosive event which occurred in the Tunguska region of Siberia on June 30, 1908 was a great meteorite or comet, two scientists at the Center for Relativity Theory at the University of Texas, A.A. Jackson IV and M.P. Ryan Jr., have explained the event as having resulted from the passage of a mini black hole through the earth. Their suggested test of the theory, by hunting through old "ships' logs" for any record of the expected air and sea shock disturbances accompanying the re-emergence of the black hole in the North Atlantic has so far been stymied by Russian refusals to provide the vital records. (Tass, the Soviet News Agency, comments: "Bourgeois capitalist Americans, in an attempt to discredit the greatness of the People's Meteorite, which fell within Mother Russia in 1908, have put forward the ludicrous suggestion that it was a black hole. That most degenerate of all western inventions . . .")

vincing case for a black hole. In all eight cases Trimble was able to conjure up a semireasonable explanation for why the dark companion was invisible without resorting to the hypothesis that it was a black hole.'' For example, he said, the unseen companion might actually be two ordinary stars, together very massive but individually not bright enough to be visible.

Among the candidates considered unlikely by Thorne and Trimble was the companion of a supergiant. Epsilon Aurigae. The latter was estimated to be 32 times more massive than the sun, and its companion was also very large—roughly 23 solar masses. Since early in this century, as noted in 1970 by Alastair Cameron, the system had been regarded as ''very mysterious.'' The two objects are circling each other once every 27.1 years at a distance comparable to that between the sun and the outermost planets (Neptune and Pluto). For 700 days of this period the larger star is partially eclipsed but, as Cameron pointed out, ''The eclipses are extremely unusual.'' The light of the larger star is only partially cut off, as though it were shining through semitransparent material. None of the observed light seems to originate in the companion, except possibly during each eclipse. Cameron, with backing from Richard Stothers, his colleague at the NASA Institute for Space Studies, proposed that the companion is a giant star that collapsed without explosively blowing off more than a small percentage of its material. The black hole would be buried inside a semitransparent cloud. Others have argued that the dust simply hides a normal star or envelopes a new star and planetary system in process of formation. But if light from the visible star passes through the cloud it seems odd that a star buried within it could not be seen at least dimly. In any case the debate on Epsilon Aurigae remains unresolved.

It was not until 1978 that a black-hole candidate as plausible as Cyg X-1 was identified. That such an object might be orbiting the supergiant star HD 152667 had been suggested in 1972 by E. N. Walker of the Royal Greenwich Observatory, based on his observations that the star was circling an unseen companion whose extremely powerful gravity pulled the star into an elongated shape. Twice, each time the waltzing partners made one full circuit, one of the supergiant's two narrow ends faced the earth and its light reached a minimum. The timing of this darkening effect, as well as Doppler shifts in spectral lines from the star, showed the rotation rate to be 7.84825 days. Walker suggested that the unseen companion might be the X-ray source Sco X-2, which also lay in the constellation Scorpius, although its position was not well known.

It was 1978 observations with the Copernicus satellite that detected the companion as a previously unknown X-ray source which Ronald S. Polidan of Princeton University, the chief experimenter, designated OAO 1653-40 (the numerals giving its celestial co-ordinates and the OAO indicating discovery by an Orbiting Astronomical Observatory). Copernicus carried British X-ray detectors as well as a Princeton telescope for obtaining ultraviolet spectra. The ultraviolet and X-ray instruments were so aligned that it could be assumed they scanned the same target.

When the new X-ray source was first observed, from April 24 to 26, the X-ray intensity fell off sharply near the end of that period as though being partially eclipsed as the source went behind the supergiant. If so, it was reasoned, and if the objects circle one another every 7.8 days, another eclipse should be expected May 27. On that date Copernicus was commanded to take a second look and, sure enough, the eclipse occurred. The X rays were not entirely cut off, but in early September a complete eclipse was recorded and was very abrupt—occurring in less than an hour. As with Cyg X-1, no rhythmic pulses were evident.

Polidan and his British coworkers (G. S. G. Pollard, P. W. Sanford, and Maureen C. Locke) estimated the mass of the visible star (known also as Hr 6283 or V861Sco) at between twenty and thirty solar masses, which implied from seven to eleven solar masses for the X-ray source. "This range," they reported, "is above the limit for a neutron star indicating that the compact secondary is a black hole."

One way to strengthen the black-hole argument would be to figure out as precisely as possible what happens when material that has been sucked out of a normal star by its black-hole companion spirals down toward the hole (as illustrated in the end papers of this book) becoming squeezed and hot enough to generate bursts of extremely high-energy X rays. If the calculations were correct, it might be possible to predict other manifestations that could be observed as confirmation of the theory. Among those tackling the problem were Kip Thorne at Cal Tech, Ruffini and John Wheeler in Princeton, Nikolai Shakura and Rashid Sunyaev in Moscow, and in England James Pringle and Martin Rees (the latter having succeeded Hoyle as head of the Institute of Theoretical Astronomy at Cambridge.)

In the 1960s Geoffrey Burbidge and others had calculated that, where a white dwarf is in tight orbit around another star, gas from the latter would not fall directly onto the dwarf but would first form a disk around it, somewhat like the rings of Saturn. This, it was widely agreed, would occur around a black hole. It was proposed that the black hole of Cyg X-1 is twelve kilometers wide (although, since space becomes highly curved in such a region, it is more realistic to speak of its circumference). As gas from the star falls toward and overtakes the rapidly orbiting hole, the gas is drawn into swirling motion around the hole. As with any such rotating system, be it a galaxy of stars or a cloud of dust and gas condensing into planets, centrifugal force flattens it into a disk, in this case a million kilometers or more in diameter, tens of thousands of kilometers thick at its outer edge, but only a few kilometers thick near the hole. In Kip Thorne's reconstruction it takes the gas a few weeks to spiral in until, in the last few hundred kilometers, it becomes heated above ten million degrees and emits the X rays observed from earth. The gas would spiral in more rapidly were it not for friction from gas in the innermost, fastest-moving part of the disk, which accelerates the outer part, keeping it in orbit longer than would be the case otherwise. This, however, steals angular momentum from the inner part of the disk, whirling around the black hole at 160,000 kilometers per second, and it plunges into the hole.

Sunyaev, in the Soviet Union, pointed out that within such a disk the infalling gas, squeezed by the extremely strong gravity of the black hole, could develop hot spots of very high temperature. Since these spots would be deep within the spinning disk, they would take only a very small fraction of a second to make one trip around the hole. While Cyg X-1 can double its X-ray brightness in one twentieth of a second, it was also known that it emits trains of X-ray flashes, like bursts of machine-gun fire. If uniformly spaced, it was proposed, these could be caused by hot spots that survive for several orbits, aiming an X-ray flash at the earth each time around. Hot spots at the inner edge of the disk would go around fastest and would emit pulses only a thousandth of a second (a millisecond) or less in length (specifically, 3.6 milliseconds for a nonrotating hole and 0.6 millisecond for a fast-spinning one).

On October 4, 1973, a NASA rocket was launched from White Sands equipped to look for such extremely short X-ray bursts, and the experimenters reported afterward: "We have evidence of the variability of Cyg X-1 on time scales down to a millisecond, consistent with turbulence in disk accretion." A train of three peaks were spaced five milliseconds apart, which Ruffini calculated was the period of the innermost stable orbit around a black hole of ten solar masses. "So strong an agreement between theoretical prediction and observed fact," he said (according to *Science*), should be accepted as verification.

The skeptics, however, have remained unconvinced and as of early 1979 the High Energy Astronomy Observatory, HEAO-1, had not detected the predicted trains of uniformly spaced X-ray flashes from Cyg X-1 or the other two candidates of similar type, Cir X-1 and GX 339-4. HEAO-1 was launched into orbit on August 12, 1977, carrying X-ray and gamma-ray detectors devised both by those believing black holes to be real (as the Naval Research Laboratory group) and those deeply skeptical (the MIT doubters). HEAO-1 has observed "shot noise" from all three sources—bursts of X-ray flashes occurring in random, shotgun fashion. A typical burst lasts about a half second, composed of millisecond flashes with no rhythmic spacing. As will be seen in Chapter 16, it is elsewhere that HEAO-1 has found the rhythmic flashes predicted for black holes.

One of the most powerful confirmations of Einstein's theory—and of black holes—would be the observation of gravitational waves. Since, according to Einstein, the gravitational field produced by a particular assemblage of matter is a curvature of space-time, any sudden change in that assemblage will generate a wave motion through space—a gravitational wave. Such ripples in space-time would move at the speed of light and would be so weak that, even if generated by a catastrophic event—the collapse of a star, or a collision between black holes—they would be very difficult to detect. The pioneer in such efforts has been Joseph Weber at the University of Maryland. For more than a decade (during which Virginia Trimble joined him in matrimony as well as astronomy) he has tried to measure, electrically, the very slight distortions that passing gravitational waves would produce in objects such as cyclinders of solid aluminum two feet in diameter and five feet long. He has gone to extraordinary lengths to eliminate other distorting effects, such as temperature changes and air movements. The cylinders were hung in a vacuum chamber that rested on shock absorbers to minimize tremors from passing trucks.

Despite all these precautions he recorded more "gravitational waves" than most theorists were prepared to accept, but at the Ninth Texas Symposium on Relativistic Astrophysics, held in Munich in 1978, Joseph H. Taylor of the University of Massachusetts reported the first indirect evidence for the radiation of gravity waves. He said orbital motion of the binary pulsar discovered four years earlier had, in that time, slowed four ten thousandths of a second—precisely as predicted for the "running down" of such a rotating, gravitationally asymmetric system through the emission of gravity waves. The pulses occur sixteen times a second and variations in that rate caused by orbital motion made possible extremely precise timing, if all sources of error were eliminated. This required taking into account such effects as variations in clock rates on the earth as the planet, in its own elliptical orbit around the sun, moves through solar gravity of varying intensity.

Efforts at direct observation of gravity waves are under way from Moscow to Peking, from Rome to California and Australia. Experimenters are using bars of sapphire crystal, cylinders magnetically suspended in space, laser beams reflected back and forth between multiple mirrors, and the tracking of spacecraft to detect "jiggles" in their separation from the earth. In all cases the purpose is to record brief distortions of local geometry as a wave passes, but the effects are probably so subtle that it may be well into the 1980s before measuring systems are sufficiently sensitive.

So far as black holes themselves are concerned, perhaps not until a new generation of theorists comes on the scene, armed with observations made from space with more sophisticated instruments, will a consensus be reached—if then. The debate is not an idle one, for its resolution would bear on the essential nature of time, matter, and space.

12
White Holes, Wormholes, and Naked Singularities

The possibility that black holes may be a reality has shaken modern physics to its foundations. It has opened up a wide range of speculation—from the proposal that we live inside one vast black hole (the universe itself) to such farfetched ideas as using small black holes for power generation or diving into a large one, slightly off center, to reach distant realms of space and time (past or future). When, in 1973, the National Academy of Sciences in Washington celebrated the five-hundredth anniversary of the birth of Copernicus, John Archibald Wheeler, the theorist who coined the term "black hole," told the assembled savants that the possibility of the universe eventually collapsing to infinite density "confronts the physics of our day with the greatest crisis in all the history of science."

Particularly appealing to science-fiction writers is the concept of "wormholes," which tunnel through the contorted space-time geometry of black holes into other universes—or emerge into our own universe at some other time and place. If a star went through such a wormhole it might, according to one hypothesis, burst forth far away, in space-time dimensions, radiating intense energy. Such "white holes," according to Igor Novikov in the Soviet Union and Yuval Ne'eman in Israel, could, for example, account for the extraordinary brilliance of quasars. Robert M. Hjellming of the United States National Radio Astronomy Observatory suggested in 1971 that through black holes and white holes matter and energy flow back and forth between our universe, dominated by matter, and a universe formed of antimatter. "The concept of white holes," he wrote, "has been invoked, using slightly different language, by Hoyle and Narlikar to identify the sites at which matter is 'created' in a steady-state cosmology."

His reference was to a series of papers by Fred Hoyle and his former student, Jayant Vishnu Narlikar, that discussed how matter could be created, through the action of an otherwise unobserved "C" field, to fill space made empty by constant expansion, providing for an eternal, unchanging ("steady state") universe. Hoyle, in a 1969 proposal, envisioned equal parts of matter and antimatter being formed in the nuclei of galaxies, most of the antimatter remaining there and the matter being ejected to form gas, dust, and stars.

Actually the idea that newly formed matter is erupting from cores of galaxies had been proposed as early as 1928 by the British astrophysicist and mathematician Sir James Jeans. While the general theory of relativity was more than a decade old, no one had yet begun to take seriously the concept of a point where density becomes infinite and time and space in effect disappear—a "singularity." But Jeans wondered—as astronomers have ever since—how the spiral arms of galaxies could be formed and remain intact despite all the motions within them. The idea that galaxies—then called spiral nebulae—were distant "island universes" far beyond our own Milky Way Galaxy had only recently become accepted. Perhaps, Jeans thought, their spiral arms are formed by streams of matter shooting out from the core of each galaxy, like water from a rotating sprinkler.

> Each failure to explain the spiral arms [he wrote] makes it more and more difficult to resist a suspicion that the spiral nebulae are the seat of types of forces entirely unknown to us, forces which may possibly express novel and unsuspected metric properties of space. The type of conjecture which presents itself, somewhat insistently, is that the centres of the nebulae are of the nature of singular points, at which matter is poured into our universe from some other and entirely extraneous spacial dimension, so that to a denizen of our universe, they appear as points at which matter is continuously being created.

It was an idea kept alive in subsequent years by Soviet-Armenian theorist Viktor A. Ambartsumian.

The wormhole concept has its roots in calculations by Einstein and his collaborator Nathan Rosen at the Institute for Advanced Study in Princeton, New Jersey. In 1935 they sought to find some sort of unifying principle linking current theories on atomic particles (such as quantum behavior) with those on electricity and on the curved space-time geometry of relativity. It was distressing to Einstein that such important aspects of our understanding of nature should seem to have no relationship to one another.

"In spite of its great success in various fields," the two men wrote in the *Physical Review*, "the present theoretical physics is still far from being able to provide a unified foundation on which the theoretical treatment of all phenomena could be based." Thus, they said, while general relativity is applicable to matter on a large scale, it has "hitherto been unable to account for the atomic structure of matter and for quantum effects." On the other hand, they continued, "we have a quantum theory, which is able to account satisfactorily for a large number of atomic and quantum phenomena but which by its very nature is unsuited to the principle of relativity."

Some theorists, Einstein and Rosen pointed out, had sought to describe atomic particles as tiny singularities—points of infinitesimal size within which the values of such characteristics as density become infinite. "This point of view, however, we cannot accept at all," they said. "For a singularity brings so much arbitrariness into the theory that it actually nullifies its laws." They proposed instead that particles consist of "bridges" between two "sheets" of relatively flat space-time geometry (the kind we live with). Traveling from one sheet to the other, the gravity—and hence curvature—becomes increasingly tight but, before closing into a singularity, they open out again to the flatness of the other sheet. It was a concept that caught the imagination of John Archibald Wheeler, then aged twenty-two and soon to join the university faculty at Princeton. It was developed by him into a novel theory of electricity.

Einstein and Rosen were not trying to build bridges between universes but rather to explain phenomena on the subatomic scale. Furthermore, they were not dealing with collapse by what would presumably be a rapidly spinning object. But in the 1960s the new school of black-hole theorists harkened back to the geometry of the "Einstein-Rosen bridge" as a clue to what might happen under more dynamic circumstances.

The possibility was explored that something—or someone—falling into a black hole born from asymmetrical collapse might escape total annihilation in its singularity and tunnel through a wormhole into some other region of space and time. Such tunneling would be through a hypothetical inner "anti-event horizon" or "hypersurface." At Birkbeck College, London, Roger Penrose and Michael Simpson sought, with the aid of a computer, to assess "event horizon" in the "wake" of the collapsing body could pass through an inner horizon to some never-never land. It appeared, they said, that such an inner horizon would be highly unstable. Furthermore, the observer, after traversing an intermediate realm of space-time, would, because of converging factors, observe "the entire history" of the region just passed, bombarded by all its radiations and tied in knots by its unbounded curvatures. Therefore, they wrote with notable understatement in the *International Journal of Theoretical Physics,* "we conclude that the projected journey of our hypothetical observer . . . looks liable to prove a dangerous undertaking. . . ." It appears, in fact, that the inner horizon would be so unstable that any wormhole would remain open only an instant, if at all.

One way that theorists attack such problems as whether or not wormholes exist is to think in terms of "world lines." All particles in the universe, in their transformations and travels through space and time, can be thought of as describing an endless history—a "world line." Such lines should go on forever and ever, either curving so as to remain within a closed universe or curving outward to infinity. The idea of a world line that ends in a singularity troubles theorists. The alternative would be a line that continues on "somewhere else" through some sort of Einstein-Rosen bridge. This puzzling problem has been likened to that concerning the inability of fleet-footed Achilles to overtake a tortoise in the famous para-

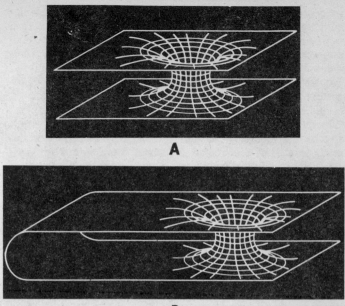

A

B

Two concepts of an Einstein-Rosen bridge in space-time geometry as applied to the problems of singularities and wormholes. In one case (A) the bridge links two separate universes. In the other (B) it joins two parts of the same universe, widely separated in space and time. What is essentially a four-dimensional concept is shown in three dimensions. (Adapted from William J. Kaufmann III, *Relativity and Cosmology*, New York: Harper & Row, 1973)

dox of Zeno, the Greek philosopher of the fifth century B.C. The tortoise was given a head start and, when Achilles reached the animal's starting point, the tortoise had advanced a short but finite distance. By the time Achilles had covered that added distance the tortoise had advanced another little bit; and so on, *ad infinitum*. Achilles could never catch up. That, of course, is not the case in the real world, and likewise, it is argued, the world line of a particle need not necessarily be halted within a black hole.

Nevertheless, few theorists are persuaded by the concept of matter tunneling between universes and most dismiss as preposterous any suggestion that an intrepid astronaut in a suitably sturdy spaceship could traverse a wormhole. Long before reaching the black hole, it is argued, the gravity gradient would pull spaceship and astronaut apart. That is, if he were approaching feet first, the gravitational field at his feet would become so much stronger than the gravity one or two meters farther from the hole that, were he made of rubber, he would be stretched to an unrecognizable shape.

Despite their doubts as to the reality of wormholes, a number of theorists consider them a fascinating theoretical challenge. They are treated as such in the textbook on gravitation written for advanced physics students by Wheeler with Kip Thorne and Charles Misner, both of whom at Princeton had profited by his remarkable qualities as a teacher. Within the infinite space-time curvature of a singularity, the three men noted, the mathematical equations of general relativity "lose their predictive power." Yet, they pointed out, according to one interpretation of those equations, a wormhole linking singularities in two universes would be "dynamic"—expanding, then contracting to break the link. As a solution to the equations, "this expanding and recontracting wormhole must be taken seriously. It is an exact solution; and it is one of the simplest of all exact solutions." But, as Thorne has put it, while wormholes have not been proved impossible, they seem highly improbable. As stated by the three authors: "There is no reason whatsoever to believe that such wormholes exist in the real universe."

A ticklish problem in assessing the possibility of black holes and wormholes is determining whether asymmetry could prevent collapse to total invisibility. Astronomical objects are never completely symmetrical. Might lopsidedness and fast rotation deform the collapse to the point where black holes could never occur or would fly apart as soon as formed? As noted by Stephen Hawking, it might be that the addition to a spherical collapsing star of a small electric charge or of a small amount of angular momentum would "completely alter the nature of the solution." Could the event horizon—that black-hole frontier through which, theoretically, no information can escape—be sufficiently distorted to permit a glimpse of what is inside—exposing a "naked singularity"?

These challenging questions have wooed the attention of

many theorists, among them James M. Bardeen at the University of Washington in Seattle, Hawking and Brandon Carter at the University of Cambridge, Werner Israel at the University of Alberta, David C. Robinson at King's College, London, and others. At the California Institute of Technology William H. Press and Saul A. Teukolsky have tried to see if a rotating black hole could settle down to a stable state. "The least pleasant possibility," they found, was that the event horizon could snap open, allowing information from within the hole to escape.

Such a naked singularity would represent a breakdown in laws that everywhere else govern natural phenomena. The rules of probability would become invalid. Causality would no longer apply and no prediction could be made as to what might come out of the hole—it could be anything (as Press has put it) "from television sets to busts of Abraham Lincoln." The experience of falling through an event horizon has been compared with death—entering "the undiscover'd country from whose bourn no traveler returns." To peek inside would be like seeing someone return from the dead.

The circumstances of collapse dealt with by Press and Teukolsky did not permit a naked singularity, but, they wrote in 1973, "one hopes that naked singularities will at some future time be ruled out by something more general than a case-by-case analysis . . . but at present they are not."

In any event the theorists seem generally agreed that asymmetrical collapse can end in a stable black hole. Even if, as Cal Tech's Kip Thorne has put it, the surface of the hole "may have a grotesque shape and may be vibrating wildly," within a fraction of a second it probably smooths out to a uniform, symmetrical shape and, once formed, cannot be destroyed.

Another challenge was to determine, by theoretical analysis, whether, when a star has collapsed to invisibility, it inevitably contracts to the infinitesimal dimensions and infinite density of a singularity. It was suspected that mathematical "overidealization" might be responsible for the prediction of so radical an end result. Perhaps the concept was a simplification derived by artificially smoothing Einstein's cosmological equations. Hawking and Roger Penrose showed, however, that if Einstein's theory is at all applicable, even without such smoothing of the equations, collapse to a singularity is inevitable.

Among the more significant developments in black-hole theory has been the apparent demonstration that the holes, spinning rapidly, can give off energy. If a star, to conserve its angular momentum, must spin a hundred times a second or more after first collapsing into a neutron star, the spin rate of a black hole would be almost unimaginable. In studying the role of spin Roy P. Kerr, a New Zealander, deduced that a black hole must have two surfaces. The outer one would be where gravity becomes so strong that it inevitably drags any matter or light in the direction of its force. Because of the spin of the hole, this force is not inward, but rather is directed circumferentially around the hole. Light emitted in this region spirals around the hole; if it is directed outward it eventually escapes the outer surface and propagates normally into space. But it turns out, as pointed out by Roger Penrose, that the energy of this light beam can exceed the energy needed to emit it. The added energy comes from the rotation of the hole. This energy-producing layer of a black hole is known as the ergosphere, from the Greek word whose derivative, "erg," defines the basic unit of energy.

The inner boundary of the ergosphere would be the "event horizon" where gravity became so strong that not even the highest-energy particles and light waves could escape. A particle falling toward this event horizon would seem to an outside observer to slow to a halt because it was entering a region where, from such an observer's perspective, gravity was bringing time to a standstill. The particle would appear to hover eternally just above the event horizon. Another process is envisioned by Stephen Hawking—that remarkable wheelchair theorist—which could cause mini black holes to "evaporate" entirely. His catch phrase for this concept is that "black holes are white hot." He cites evidence that space, far from being empty, is alive with activity in which pairs of particles—one of matter and one of antimatter—are constantly being formed only to meet again and annihilate one another. They are a consequence of quantum behavior on the subatomic level and are what physicists call "virtual" particles that, by definition, cannot be observed directly, even though there is indirect evidence for their existence.

As noted earlier, when a particle of matter meets one of antimatter—for example, an electron and its antimatter twin, a positron—they vanish in a flash of energy (an extremely energetic gamma ray). The reverse process also occurs, as routinely observed in high-energy physics experiments. When a very energetic collision occurs, some of the energy emerges (for an unobservable instant) as a gamma ray, which then materializes into an electron-positron pair.

According to Hawking the enormously powerful gravity around a black hole would generate such pair production in abundance. In any other situation the pairs would quickly meet one another and annihilate, turning back into energy, and there would be no net loss of energy to the system. But near a black hole one of the particles may vanish into the hole, leaving its sister particle free to fly off.

If the particle that falls into the hole is a positron, the process can be viewed as the escape of an electron from the hole—previously considered an impossibility. This is the consequence of a concept developed by Richard P. Feynman, whose contributions to physics theory won him a Nobel Prize. In his view, time, as it applies to antimatter such as positrons, may flow backward. What looks like a positron falling into the hole can therefore be viewed as an electron tunneling out of the hole backward in time. Its interaction with the hole's gravity then reverses its time direction and converts it into an electron free to fly off into the universe. On the other hand, if it appears that an electron has fallen into the hole, in the Feynman concept this can be interpreted as a positron escaping by motion backward in time until, just outside the hole, its interaction with the powerful gravity of the hole reverses its time direction and it flies off toward the future. In this admittedly mind-boggling manner a black hole can "evaporate" both positrons and electrons. To some theorists this demonstration that particles (governed by quantum theory) can "tunnel" out of a black hole "governed by relativity) is the most important unifying development since Einstein defined the mass-energy relationship.

The region through which a particle must tunnel is thin-

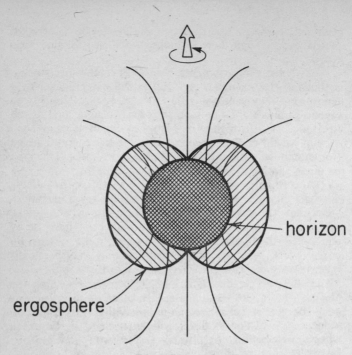

BLACK HOLE

A spinning black hole has two surfaces: an outer one from which light, dragged to higher energy by the gravity of the rotating hole, can escape, and the inner "event horizon," through which nothing can escape. Between is the "ergosphere." *(Princeton University).* The end papers of this book show an artist's conception of a black hole and its companion star, done by NASA's Goddard Space Flight Center.

ner for a little black hole—one of small mass—than for a big one. Consequently, small holes radiate away their energy much faster than big ones, thus depleting their mass, which in turn makes them radiate even more intensely. They become "white hot" in a runaway process ending in a catastrophic explosion. This would apply in particular to the mini black holes that Hawking said might remain from the Big Bang (such as the one blamed for the Tunguska explosion). The tiniest, no larger than a nuclear particle, but weighing as much as a mountain, might have radiated away so much energy that it "evaporated" explosively.

George F. Chapline of the Lawrence Livermore Laboratory in California has suggested that black holes of all sizes may be so common that a region the size of our solar system might contain several tiny ones. "Density fluctuations on very large mass scales were certainly present in the early universe," he wrote in a 1975 issue of *Nature*, "as is evident from the irregular distribution of galaxies in the sky." Thus the exploding universe was not homogeneous and the expansion of some parts squeezed other parts, forming in his view a complete range of black holes from ones no larger than an atomic particle to those equivalent in mass to millions of stars.

Theorists have calculated, from the amount of deuterium relative to helium formed from three to four minutes after the start of the Big Bang (as will be discussed in a later chapter), that there were insufficient nuclear particles (nucleons) in the exploding cloud to provide the gravity needed to prevent eternal expansion of the universe. Chapline proposed, however, that during those crucial early minutes "most of the matter in the universe must have existed in some other form than free nucleons—in other words, black holes."

In the spring of 1977, to see if tiny black holes are still exploding in the manner predicted by Hawking, a multimirror reflector of the Smithsonian Astrophysical Observatory atop Mount Hopkins in Arizona was used in tandem with the mirror system of a solar furnace at White Sands, New Mexico—a mirror array nine meters on a side. The Smithsonian's multimirror reflector (ten meters wide and not to be confused with its innovative multimirror telescope at the same site), is designed to record the effects of extremely high-energy cosmic rays (gamma rays) hitting the upper atmosphere, rather than to observe distant objects. Such a gamma ray would strike with sufficient energy to produce a cascade of electrons moving so fast that each generates a flash of light (Cerenkov radiation). Hawking's predicted bursts of gamma rays would result in a similar electron shower, but of vast extent. If the effect were recorded simultaneously by Mount Hopkins and by the array of solar-furnace mirrors 400 kilometers away, this would indicate that the event was not a local cosmic-ray event and would lend support to the hypothesis.

The search, sponsored by the Smithsonian Research Foundation and the National Research Council of Ireland, was said to be the first time so much mirror area had been used for an astronomical experiment. However, during twenty-two and a half hours of observing on ten moonless nights, no coincident flashes were recorded. This could mean

1. that miniholes do not exist,
2. that they are much rarer than some had hoped, or
3. that they do not explode in the manner predicted.

This last possibility had been proposed in 1974 by Paul C. W. Davies and John G. Taylor of King's College, London. According to their hypothesis, a minihole would emit a flash when formed but would not die explosively.

In discussing the possible existence of miniholes within the solar system Chapline said this "might have considerable economic significance because small black holes would be very useful as power sources." His reference was to a proposal made the previous year by three of his colleagues at Livermore—among the most far-out of all those relating to black holes. The Lawrence Livermore Laboratory, operated for the federal government by the University of California, does energy and weapons-related research, much of the effort being aimed at harnessing the energy source of stars—the fusion of hydrogen into helium. The trio of researchers—John Nuckolls, Thomas Weaver, and Lowell Wood—proposed a way to extract energy from nearby mini black holes (some perhaps orbiting the earth). The miniholes would be so tiny their gravitational fields would produce an effect out to only a few hundred meters. In fusion fuel, such as the heavy forms of hydrogen (deuterium and tritium), were fired toward such a minihole, the fuel would become enormously compressed just before vanishing into the hole—squeezed sufficiently to cause those atoms to fuse into helium, as occurs in the core of the sun. The released energy could be captured and relayed to earth by a space station that was orbiting near the little black hole and firing fusion fuel into it. While this, the three men said, may appear to be "enormously more difficult" than more conventional approaches to fusion, the problems "may have been overestimated." When Wood presented this scheme on behalf of his colleagues at a meeting of the New York Academy of Sciences, it evoked gasps and chuckles. It was, he insisted, a serious proposal.

Two years earlier Press and Teukolsky, in a somewhat tongue-in-cheek offering in *Nature*, had discussed the possibility of a "black-hole bomb" and, alternately, of ways to extract energy from such a system. The essence of the concept was to enclose the hole in a spherical mirror at a suitable distance. For a black hole equal in mass to the sun the distance might be about one thousand kilometers, they said. The mirror would reflect low-frequency radio waves back toward the hole where, in essence, the waves would pick up rotational energy from the hole. If the mirror were completely spherical the pressure of this amplified radiation would finally burst it—the "black-hole bomb." If there were an opening in the mirror, escaping radiation, it was proposed, could be "rectified and used as a source of electric power" (much as radiation escapes through an aperture in the mirror system of a laser).

The first proposal for using black holes as energy sources seems to have been that in the early 1970s of Misner, Thorne, and Wheeler. In their book on gravitation they described an advanced civilization that has enclosed a black hole within a rigid sphere on top of which is a "huge city." The hole's gravity holds this city firmly on its spherical support structure.

Each day a million tons of garbage from the city are collected and loaded onto shuttle vehicles that are dropped toward the hole. As the vehicles spiral ever more rapidly in toward the hole, the garbage is ejected and vanishes into the hole. The vehicle, however, recoils from the ejection and flies back out, having meanwhile gained enormous energy and velocity transferred to it by the hole during the vehicle's spiraling flight. The hole, at the same time, gains energy from the infalling garbage. The shuttle on its return flight is captured by the city in a manner that transfers its "huge kinetic energy" to a flywheel. The latter drives an electric plant supplying the entire city.

Since, in transferring energy to the vehicle, the black

hole forfeits mass, the three authors wrote, "not only can the inhabitants of the city use the black hole to convert the entire rest mass of their garbage into kinetic energy of the vehicle, and hence into electric power, but they can also convert some of the mass of the black hole into electric power!"

The presentation, complete with supporting equations, was as much an exercise for the student as a proposed solution for energy problems of the distant future, but a similar idea has been far more extensively developed by Paul Davies, the King's College physicist, in his book *The Runaway Universe*. He envisions a universe that has almost entirely come under intelligent control, deriving its energy from black holes instead of stars and thus able to survive for "a billion billion" years after all the stars have burned out. "Using rotating black holes," he says, "supertechnologies could prolong their lifetimes so much they would become the dominant activity of the Universe for a longer period than the stars." After a billion billion years of black-hole manipulation, the epoch of starstudded skies would linger only as a dim memory. He proposes that a civilization seeking to prolong its life "set about collecting, storing and coalescing" mini black holes. It need hardly be pointed out, however, that to date no vehicles have been launched on such missions—at least from this planet.

Does the truth regarding what happens inside black holes lie forever beyond our reach, denied us by what Penrose refers to as "cosmic censorship"? If, as the Big Bang suggests, the universe we know was born from a singularity—with infinite density, infinite pressure, infinite space-time curvature— some theorists contend that, in reconstructing from the clues at hand what happened at the birth of the universe, we are looking back close to the singularity itself. In this sense we are inside of what can be regarded as a black hole—the whole universe. In the words of Kip Thorne: "Our universe seems to have exploded to create this space and time, and we are trapped inside its gravitational radius. No light can escape from the universe."

But it cannot be argued that in looking back toward the Big Bang we are actually viewing a naked singularity with all of its outrageous properties, much less that we can look through the singularity to what came before. It may well be that "cosmic censorship" will forever—and in all universes—prevent "everyone" from doing so. Whence did we come? And where will we go? Such questions, in the grand sense, have long been relegated to theologians, and now some physicists are saying that nature has placed the answers intrinsically beyond our reach.

Chandrasekhar, who in 1930 first defined the limiting mass of a white dwarf and later became convinced that black holes probably exist, recalled many years later a parable from his childhood in India that described an information barrier in the lives of dragonflies much like that of the black-hole "event horizon." In their larval stage dragonflies are ugly, predacious creatures that live on the bottom of ponds:

> A constant source of mystery for these larvae [said Chandra in a 1972 lecture at Oxford University] was what happens to them, when on reaching the stage of chrysalis, they pass through the surface of the pond never to return. And each larva, as it approaches the chrysalis stage and feels compelled to rise to the surface of the pond, promises to return and tell those that remain behind what really happens, and confirm or deny a rumour attributed to a frog that when a larva emerges on the other side of their world it becomes a marvelous creature with a long, slender body and iridescent wings. But on emerging from the surface of the pond as a fully formed dragonfly, it is unable to penetrate the surface no matter how much it tries and how long it hovers. And the history books of the larvae do not record any instance of one of them returning to tell them what happens to it when it crosses the dome of their world.

Even if it were possible to bypass an "event horizon," looking inside might reveal nothing of the black hole's history, for within it "all is forgotten." What goes into the hole loses all characteristics except mass electrical charge, and angular momentum. Take, for example, an astronaut unfortunate enough to fall victim to the gravity of a black hole, drawing him inexorably toward its singularity. First his body and then the molecules of which it is formed are torn apart. So are the atoms that formed those molecules and, after that, the "elementary" particles from which those atoms were composed, although some theorists wonder what new effects manifest themselves when the space-time curvature tightens to the dimensions of an atomic particle. Perhaps, they think, quantum effects then play the dominant role—at last bringing quantum theory and general relativity together, as Einstein had hoped.

In any case, once a black hole has formed, it has no "memory" of what it once was. It could have been a giant star (formed of matter or of antimatter) that collapsed. It could have been in any shape or form. As Martin Rees, Remo Ruffini, and John Wheeler have pointed out in their book on the subject, it could have been simply a cloud of energy that—being equivalent to mass—collapsed upon itself. Or it could have been a combination of such starting materials. All that remains is its mass, electric charge, and angular momentum.

If a whole universe collapsed to a singularity. Wheeler has pointed out that even those otherwise imperishable features would lose their meaning since they are properties relative to "something else" that would no longer exist. If, from this singularity, a new universe was born, there would be no reason why its physical laws, determining the nature and behavior of atoms, and such physical "constants" as the strength of gravity, should be the same as in this universe since, having been "squeezed through a knot-hole" and drastically "reprocessed," this new universe would have no memory of what had ruled its predecessor. The rules, Wheeler said, "are freshly given for each fresh cycle of expansion of the universe." If the rules were much different from those in the present universe, the physics, as noted by Brandon Carter, might not be suitable for the evolution of stars like the sun or the creation of those substances essential to life. Our chemical ancestry lies within the cores of stars and in the explosions of supernovas in which all the elements except hydrogen and helium are believed to have been synthesized. Other universes with different rules and different chemistry accordingly might be lifeless.

The principle that all memory, all rules, are lost in a black hole has been summarized by Wheeler with the theorem: "A black hole has no hair." But assuming our universe was born from a singularity, it seems hard to understand how all the diversity we gaze upon—galaxies, stars, planets, flowers, and people—could have evolved from such a starting point. Consequently some theorists, such as Kip Thorne, believe that if the present universe erupted from a singularity—in effect from a "white hole" of cataclysmic scope—some departures from complete uniformity may have survived from "before." One line of attack on the problem is to survey the temperature of the Big Bang glow in all directions and see if there are hot spots and cold spots or whether it is completely uniform. Such measurements, made with balloons and high-flying aircraft, have to date shown no evidence of nonuniformity in the Big Bang, although they imply that the earth and the galaxy are flying through the residual Big Bang radiation at high velocity. Such measurements, however, are difficult and are not yet definitive.

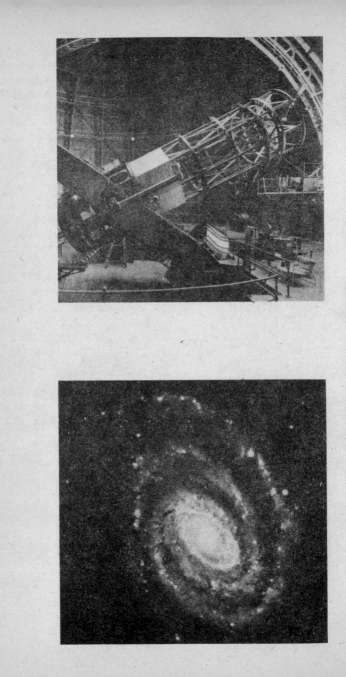

13

The Universe—Open, Shut, or According to Hoyle

Either the universe is destined to expand forever, its clusters of galaxies becoming more and more dispersed, its stars collapsed to frigid black cinders, their planets lifeless, or there is enough mass—and hence gravity—within the universe ultimately to reverse the expansion, initiating its collapse to vanish into a singularity or, perhaps, rebound to form a new universe. In the last-named case the universe may prove to be oscillating in an eternal cycle of birth, death, and rebirth in which, some believe, time may reverse its direction each time collapse begins—or at the "bounce" that forms each new universe. The only alternative to these possibilities is some exotic form of steady-state situation such as one proposed recently by Fred Hoyle.

The problem of the universe's destiny is closely linked to the question of whether or not black holes exist and, if so, what happens inside of them. In watching the expansion of the universe, as John Wheeler has pointed out, we are witnessing gravitational collapse in reverse. This, he wrote, "is evidenced not by theory alone, but also by observation; and not by one observation, but by observations many in number and carried out by astronomers of unsurpassed ability and integrity." If the expansion is destined to reverse itself, he sees the black hole as a preview of what is to come.

Such a universe is "closed." The curvature of its space, in the four dimensions of space-time, is inward, like the lines on a spherical surface. In an open universe, whose expansion is being slowed by its own gravity but will never stop, the lines curve outward (hyperbolically), like those drawn on a saddle.

It is somewhat paradoxical that a closed universe, while finite, has no "edge." Its unbounded space can be likened, in a three-dimensional sense, to the surface of the earth, whose area is finite, but which (contrary to the belief of the ancients) has no edge. It follows from the same reasoning that immediately after the Big Bang the nature of the universe was already determined—either it was infinite, in the geometric sense (even though it was not yet greatly dispersed), or it was finite. In the first case particles ejected by the explosion would travel curved lines in space-time, permitting them to continue outward forever. In the latter case their paths would ultimately curve back and return.

Such curvacious geometries had their roots in efforts by mathematicians of the eighteenth and nineteenth centuries to prove Euclid's axiom regarding parallel lines. Generation after generation was unable to show that such lines, extended to great distance, would never meet. The Hungarian mathematician Farkas Bolyai wrote his son, János: "I entreat you, leave the science of parallels alone. . . . I have traveled past all reefs of this infernal Dead Sea and have always come back with a broken mast and torn sail." The son, ignoring this advice, helped give birth to non-Euclidean geometry, as did his father's more famous friend Carl Friedrich Gauss and the Russian Nikolai Ivanovich Lobachevsky. But it was Georg Friedrich Bernhard Riemann of Germany who, more convincingly than anyone, demonstrated that "straight" lines, extended to nearly infinite length, can meet and that "space" can be curved. This set the stage for Einstein's concept of curvature under gravitational control and finally for Aleksandr Friedmann, a Russian meteorologist, to develop in the 1920s the mathematical tools needed by a new generation of cosmologists to assess whether or not the universe is open or closed. His formulations, for the first time, showed how the critical element in that regard (the slowing of expansion) depends on a combination of two factors seemingly amenable to determination: the present rate at which the universe is expanding and its average density.

To a certain extent Friedmann also anticipated the Big Bang concept. In fact, the belief grew that, to avoid collapse, the universe had to be expanding. Previously the concept, based on Newton's physics, had been of an infinite universe whose constituents, gravitationally, were tugging at one another uniformly in all directions. If the universe were finite, it was argued, it would long since have been drawn together and collapsed.

The new formulations of space and time undermined this argument. Furthermore, Einstein himself believed that the universe could not be infinite. His reasoning derived from arguments advanced a century ago by Ernst Mach, a physicist-philosopher whose views deeply influenced Einstein and many others (his name is used as a unit of speed, Mach 1 being a speed equal to that of sound in the medium through which a craft is flying).

Among Mach's many concerns was the nature of inertia—the property of matter that resists acceleration. If a rowboat is at rest it takes a hefty push to set it in motion, even though little friction need be overcome. Once it is moving it takes a comparable force to stop it. The same applies to a thrown ball. Likewise a spinning gyroscope during a twenty-four-hour period appears to rotate because its orientation remains fixed, not relative to the earth (which in that period has made one revolution) but to the universe as a whole. Thus objects, whether seemingly at rest or in motion, resist any effort to change their status relative to the rest of the universe. The effect seems related to gravity in that the mass of an object not only defines its inertia but also determines its response to gravity. The origin of inertia, in Mach's view, was an interaction between an object and all other material in the universe. "The idea that Mach expressed," wrote Einstein in *The Meaning of Relativity,* "that inertia depends upon the mutual action of bodies . . . corresponds only to a finite universe, bounded in space, and not to a quasi-Euclidean, infinite universe."

What, then, prevented gravity from pulling all parts of the universe together in a catastrophic collapse? To answer that question Einstein introduced into his equations a "cosmological term"—in effect a form of antigravity to balance gravity sufficiently for the universe to be static. When observations showed that the universe is expanding—that it is the energy of the Big Bang that fights against its collapse—he realized that, as he put it, the cosmological term was "gravely detrimental to the formal beauty" of his theory, and withdrew it.

While few now question the Big Bang cosmology, the idea of an oscillating universe that repeatedly expands, collapses to great density, then explodes again has for many a special attraction. They find eternal expansion a grim prospect, even though the expansion might eventually slow toward a halt (an almost-steady-state situation), for the stars would be burned out and the skies forever dark. On the other hand, according to Robert Dicke, the Princeton physicist and cosmologist, it is "philosophically appealing" for the universe to be cyclic, periodically contracting and reprocessing its material through an extremely hot phase. "In this way," he has written, "we avoid the necessity for arbitrarily assuming complex initial conditions for the 'start' of the universe" a few billion years ago.

It is clear, Dicke said, that in such a cycle the stuff of the previous universe must be "decomposed to pure hydrogen" in the final stages of its collapse, destroying all the more complex elements. The basis for this was the well-established fact that the oldest stars are deficient in the heavier elements.

Furthermore, theoretical physicists are troubled by the idea of an open universe that was born "from nothing." It would seem to violate conservation laws that are considered inviolable. For example, mass can be converted to energy and vice versa, but neither mass nor energy can be created from nothing nor destroyed. They are "conserved." Furthermore, when energy is converted to mass, it is well established that the newly created particles appear in pairs—one of matter and one of antimatter. Never is energy converted into individual electrons, positrons, or other such lightweight particles (leptons), and the same rule applies to the heavier particles (baryons) such as protons and antiprotons. Thus net populations of light and of heavy particles (leptons and baryons) are conserved. (In a nuclear explosion there is, essentially, no conversion of such particles into energy, the latter being derived from the "nuclear glue"—the binding energy—that is left over from the reaction.) The number of baryons or leptons remaining after all matter-antimatter pairs have annihilated one another or been otherwise accounted for must remain the same as before the reaction, be it in a laboratory experiment or—presumably—in the universe itself. Cosmologists speak of such a residual population for the entire universe as its "baryon number," and they would be happiest if it were zero. The demonstration that energy conversion into matter always produces equal numbers of particles and antiparticles implies a basic symmetry in nature that, as noted by Fred Hoyle, would seem to call for the existence in the universe of equal amounts of each kind of matter. Where matter prevails time would flow "forward," whereas in regions dominated by antimatter the flow of time might be in the opposite direction. At one of the Texas meetings on astrophysics Englebert L. Schucking, an originator of those conferences, proposed that our universe is linked to another universe formed of antimatter in which time runs backward.

In the 1950s Hoyle and Geoffrey Burbidge explored the possibility that there might be as much antimatter as ordinary matter in our universe—"antistars" and "antigalaxies," for example. If the two forms both occurred in the Milky Way Galaxy, they reasoned, the matter-antimatter encounters and annihilation would create violent motion and heating within the interstellar gas far beyond anything that can be seen. At first it seemed that there might be considerable antimatter adrift between the galaxies, but such atoms would inevitably meet atoms of matter from time to time, producing gamma-ray flashes. The effect would be a gamma-ray glow from all parts of the sky that has not been observed.

This universe seems, therefore, to be formed predominantly of matter. Was its creation in violation of the conservation laws? Or, as Dicke put it, were the particles forming it left over from its last incarnation? "The universe," he wrote, "appears to be both complex and unsymmetrical, containing enormous blocks of magnetic field of great antiquity and composed of matter instead of antimatter. It seemed evident that such a complex structure must have had a long and continuing existence with expansion followed by contraction."

An originator of the opposing view that the universe derived from one great moment of creation was Abbé Georges Lemaître, who set the stage for Gamow's more detailed description of the Big Bang explosion. Lemaître was originally trained as a civil engineer and served as an artillery officer in the Belgian Army during World War I. After the war he studied for the priesthood, was ordained in 1923, but almost immediately began studying astrophysics and astronomy, first at the University of Cambridge in England and then at the Massachusetts Institute of Technology. In 1927, upon becoming professor of astrophysics at the University of Louvain in Belgium, he set forth his view that the universe originated in a primordial "atom" whose explosion gave birth to the galaxies and stars.

In 1961, five years before his death, Abbé Lemaître was on hand at the General Assembly of the International Astronomical Union in Berkeley, California, wearing (according to my recollection) a monkish robe girded with a ropelike cord. With one finger laid on the side of his nose, as though telling a sly joke, he noted that he was sometimes accused of inventing a cosmology that demanded an act of divine creation. That, he said, was not his purpose. But, as has been noted by Hannes Alfvén, the Swedish astrophysicist and Nobel Laureate, the concept was attractive to Lemaître "because it gave justification to the creation *ex nihilo* [from nothing], which St. Thomas had introduced as a credo. To many other scientists [Alfvén continued] it was more of an embarrassment because God is very seldom mentioned in ordinary scientific literature."

In this connection it has been pointed out that hecklers once asked St. Augustine of Hippo, one of the most important early Christian philosophers, what God was doing during the timeless period before creating the world. His alleged reply was that God was creating hell for those who asked such questions.

One way to escape the problem of "a beginning" was a steady-state universe that is eternal, unchanging—yet always expanding. To keep it thus, new atoms, new stars, new galaxies must constantly be created to fill the gaps formed by expansion. While elements of such a cosmology—namely, the continuous creation of matter—had been proposed twenty years earlier by Sir James Jeans, as noted in the previous chapter, it was first fully formulated in 1958 by Hermann Bondi, Thomas Gold, and (in a separate paper) Fred Hoyle.

Their collaboration came about through a succession of

remarkable coincidences. Bondi and Gold had both been born, a year apart, in Vienna. While their parents apparently knew each other, the young men at that time did not. The Gold family left and settled in England before Hitler took over Austria in 1938. The Bondis departed Vienna immediately after the German *Anschluss,* moving to New York. Both sons enrolled at the University of Cambridge, but did not become acquainted until war broke out and they were interned by the British as enemy aliens. According to Gold, "We found ourselves sleeping side by side on a concrete floor, with no blankets." They were finally moved to an internment camp in Canada, where the two men joined with others in forming a sort of university.

When their loyalty to the anti-Nazi cause was finally accepted, they were allowed to return to Cambridge and, having demonstrated their scientific brilliance, were recruited by the Royal Navy for secret radar research. As Bondi puts it: First they were behind barbed wire as enemy aliens, then, soon thereafter, as participants in top-secret work. They ended up at an Admiralty research center at Whitley, where their section leader was a man named Fred Hoyle. During the day they worked on radar development. In the evenings they discussed astronomy, physics, and cosmology.

It was to be a long-lasting association, but Gold eventually moved to the United States and became director of the Center for Radiophysics and Space Research at Cornell University. Bondi (Sir Hermann) became chief scientific adviser to the British Ministry of Defence. Hoyle (Sir Fred), who achieved popular fame as a writer of science fiction and who even composed the libretto for an opera, *The Alchemy of Love,* headed the Institute of Theoretical Astronomy at Cambridge from 1967 to 1973, then began dividing his time between Britain and California. To astrophysicists probably his most significant theoretical contributions, made soon after World War II, concerned the origin of the elements (those heavier than hydrogen and helium), particularly in supernova explosions, leading to a memorable collaboration with William Fowler and the Burbidges (Margaret and Geoffrey).

Nevertheless, he continued to apply his fertile imagination to cosmology, and when observations—notably those of the microwave glow attributed to the Big Bang fireball—

made the steady-state hypothesis seem unlikely, he and his associate, Narlikar, devised other concepts, some of which still had elements of perpetuity. As noted earlier, Hoyle's contribution to the original steady-state concept was of an unobserved C (for "creation") field. This, he believed, was necessary to get around the conservation law, which forbids creation of the constituents of matter, such as protons, neutrons, and electrons (more precisely baryons and leptons), from "nothing."

In 1975 he proposed a new kind of steady-state situation in which yet another field—the "mass field"—is in control. At any one point in space and time this field, he said, determines the mass of every atom. In some parts of the cosmos—including what we see as our universe—the field is positive and the mass of every particle has been increasing since the time when our observations have led us to believe a Big Bang occurred. The cosmos, according to Hoyle and Narlikar, may in fact be formed of many universes, some with positive and some with negative mass fields. At the space-time boundary between two such universes the value of the field is zero.

This would put a completely different interpretation on the observed shifting of wavelengths of light from distant galaxies toward the red end of the spectrum—the famous "red shift" that has been almost universally accepted as evidence that the universe is expanding. If, as time passes, atoms become more massive, Hoyle reasoned, the wavelengths of their emissions will become shorter (bluer). Looking far out into space and back in time, the reverse will be observed. The farther we look the redder the light from a particular type of atom will appear.

In other words, Hoyle said, the universe is not expanding at all. (Only a few other rebels have questioned the expansion, arguing, for example, that as light travels long distances it somehow becomes "tired" and its wavelengths lengthen.) "The usual mysteries concerning the so-called origin of the Universe," he wrote, "begin now to dissolve."

Hoyle tried to explain the microwave glow not as the residue of a primordial fireball, but as starlight from an earlier existence of this universe that leaked through the space-time boundary when the mass field dropped to zero. Otherwise, he asked, "why is the energy of the microwave radiation observed to be so similar to that of the light generated by the stars of the galaxies?'' While matter, having lost all gravity as it crossed the boundary, would lose much of its cohesion, Hoyle said ''islands of high density'' may get through. One of the most troublesome problems in picturing a universe born from a singularity—or from a homogeneous, uniformly expanding fireball—is how it became so diverse, forming galaxies, clusters of galaxies, stars, planets, and people. But, according to Hoyle, ''it seems possible that many stars on our side [of the space-time boundary] are fossil relics of stars that existed formerly on the 'other side.' ''

In a *Nature* commentary Paul Davies (the physicist who believes black holes could become the ultimate energy source) noted that many of his colleagues ''may find this imaginative new picture a comforting circumvention of the unpleasantness of the initial singularity.'' Nevertheless, he said, it raised new problems. If, for example, starlight sneaks across the boundary from the last universe and this has been going on *ad infinitum,* why has such starlight not accumulated to unlimited brilliance?

One aspect of Hoyle's earlier concept of a C field is that (somewhat like Einstein's cosmological term) it would prevent collapse to a singularity (although it would not rule out black holes in the sense of sufficient collapse to become invisible). To avoid departure from the symmetry that seems characteristic of basic phenomena (such as the equal distribution of positive and negative electric charges) there must be reservoirs of negative energy whose ''repulsive nature,'' said Hoyle's associate Narlikar, ''will be sufficient to prevent the singularity. This conjecture,'' he continued, ''should apply to finite collapsing objects as well as to the universe.'' The object, according to the two theorists, ''bounces'' at a minimum radius ''which can lie *inside* the usual Schwarzchild radius.''

It was this idea of ''bounce'' that Hoyle contributed to his collaboration with Fowler and the Burbidges in their efforts to explain quasars and other explosive processes. Before he devised the modified steady-state concept based on changing mass fields Hoyle also saw bounce as an explanation for an oscillating universe: ''The universe alternately expands and contracts. Gravitation causes the reversal from expansion to contraction, while the new field causes the rebound from contraction to expansion." One problem, Hoyle recognized, was whether such a "bouncing" universe would gradually lose its bounce and settle into a steady state.

There are widely divergent ideas as to how dense and small the universe would become before its rebound. On the one hand, Alfvén has proposed a billion light-years as the smallest possible contraction. He dismisses as ridiculous the idea of a singularity—a universe, as he puts it, "smaller than the head of a pin." His cosmology is derived from that of Oskar Klein, a fellow Swede and physicist, who saw the universe formed equally of matter and antimatter. "It seems logically unsatisfactory," said Alfvén, that cosmologies should assume that matter predominates. When the universe—or, as Klein called it, the metagalaxy—in an earlier cycle collapsed sufficiently for the matter and antimatter to interact, this generated the radiation pressure that blew it apart. A relatively low density would be sufficient for this effect and, Alfvén said, localized matter-antimatter encounters could account for quasars and the like.

As noted earlier, the gamma-ray glow to be expected from widespread matter-antimatter encounters has not been observed, and it is also difficult to see how such a limited collapse and rebound could produce a fireball hot enough to "melt down" the dying universe, leaving only its hydrogen. It is now widely accepted that pure hydrogen was the starting material of the present universe. Robert Dicke has proposed that a fireball diameter of "a few light-years" would be adequate for this purpose. John Wheeler, his longtime colleague at Princeton, however, doubts that in the collapse of stars (and, by implication, universes) any bounce can occur, at least once a black hole has begun to form.

It is no coincidence that Wheeler, Zeldovich, and others who have analyzed the problem of bounce (as it relates to the universe, to supernovas, and even to crushing fuel pellets to produce the fusion that makes stars shine) had been concerned with the physics of nuclear explosions. A challenge in designing an effective atomic blast is to make sure the fuel is not blown apart before a significant amount of atom splitting has occurred. In any such explosion—be it in a star, a bomb, or fuel pellet—compression converges on the center from all sides, generating a rebound shock wave that radiates out to the surface. This can blast off the outer layer—as in a supernova. But then a wave of lowered density returns to the center and if, in the case of gravitational collapse, it reaches there soon enough, according to Wheeler and his colleagues, it can prevent the formation of a black hole. They concluded, however, that the collapse of a star differs from man-made nuclear explosions in that, once even a small part of the core has become a black hole, nothing can reverse the process. Thus the collapse of a massive star to a black hole, unlike formation of the Crab Nebula and its central neutron star, does not generate a great supernova explosion because there is "nothing" from which to rebound. As Wheeler put it in 1967:

> The matter of the core pours torrentially in from all directions like a thousand Niagara Falls on its way down from the original dimension to ever smaller sizes. . . . In less than a tenth of a second the collapse is complete. No core is left to serve as a dynamite charge at the center of the star. No punch arises to drive out into space the remainder of the star.

In Wheeler's view the universe, in its final stage, would collapse to a scale so tiny that the indeterminacy (as well as the behavior) that governs atoms would come into play. What happens is unpredictable: "There is no unique history that one can ascribe to the universe. Instead there is a certain probability of this, that, and the other history of the universe."

These arguments as to what happens in catastrophic situations far removed from laboratory tests recall one conducted in deepest secrecy where many pioneers of the atomic age became involved, including Hans Bethe, Arthur Compton, Enrico Fermi, and Eugene Wigner—all Nobel Prize winners in physics—as well as Robert Oppenheimer, Edward Teller, and Leo Szilard. During development of the first atomic bomb there was a concern that a nuclear explosion might generate enough heat and pressure in the air to start a self-sustaining chain reaction, consuming the entire atmosphere. The nuclei of nitrogen atoms, the most abundant constituent of air, might, it was feared, fuse into silicon, which would combine with the other primary gas of the atmosphere, oxygen, to form silicon dioxide (quartz), the chief ingredient of sand. In other words, the earth's entire atmosphere would become ignited, turn into sand, and fall to earth.

While the debate among physicists responsible for the bomb was secret, relevant documents have become available. Compton, who headed the Chicago laboratory where achievement of the first chain reaction cleared the way for the project, was later quoted as saying: "It would be the ultimate catastrophe. Better to accept the slavery of the Nazis than to run the chance of drawing the final curtain on mankind." Although the physicists became convinced that a chain reaction in the sky was impossible, in the moments before the first test explosion in 1945 Fermi, to relieve the tension, is said to have offered to bet with his colleagues on whether or not the sky would ignite and collapse. General Leslie R. Groves, the project director, reportedly was not amused.

When, in 1975, I discussed this history with Wigner, he said calculations showing the chain reaction to be impossible were, in a sense, regrettable. Had the danger proved real, the development of nuclear weapons might never have come to pass.

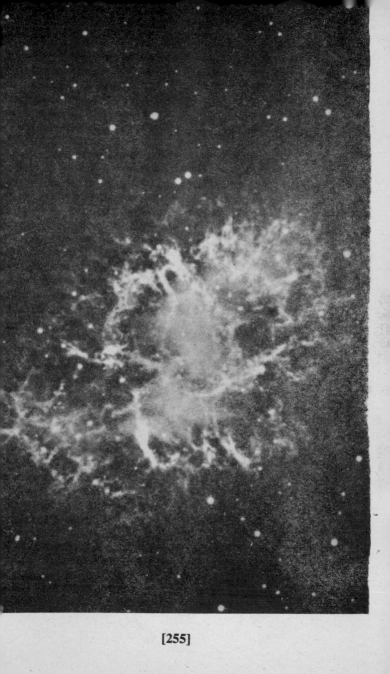

14

The Arrow of Time

The most provocative predictions concerning the future of the universe relate to the arrow of time. From our lifelong observations we are so convinced that time always flows in the same direction that it becomes very difficult to conceive of any other situation. Yet every physicist knows that, on the atomic level, a sequence of events can flow with equal ease in either direction. Take, for example, an interaction in which an atom at a high-energy state emits a light wave or radio wave, dropping to a lower state, whereupon that wave is absorbed by another atom, raising that one to a higher energy level. If, so to speak, a motion picture were made of this interaction, there would be no way, from the sequence of events, to determine in which direction the film should be run. In this sense time, on the atomic level, is reversible.

This, however, is only true in an isolated, idealized situation. In the real world the emitted wave might never hit another atom and could sail off on an endless journey. Where many atoms and many emissions are involved, as in all the mechanical devices of our world, such losses occur. Automobiles, steam engines, and the like are never 100 per cent efficient. One of the main challenges of nineteenth-century engineering was to produce steam engines that were as efficient as possible, and it became obvious to some of those working on the problem, notably the German physicist Rudolf Clausius, that a form of intrinsic energy dissipation sets irreducible limits on such efficiency. It soon came to be recognized as the "second law of thermodynamics," which

In large-scale (as opposed to atomic) events, such as pouring milk from a pitcher, time is irreversible. It is easy to tell which sequence of pictures is the correct one.

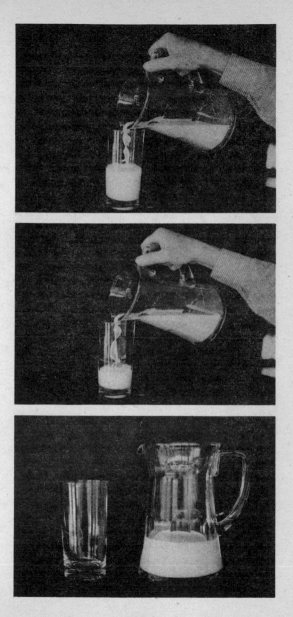

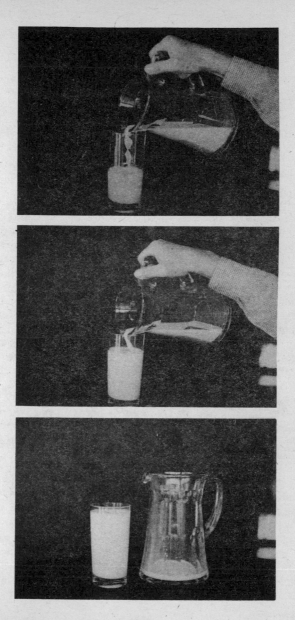

states that the entropy (or disorder) in a system, if it changes at all, must always increase.

Sir Arthur Eddington, who focused much attention on problems relating to entropy and the arrow of time, likened the situation to shuffling a deck of cards that had come from the manufacturer sorted by suits. Once shuffled, no reasonable amount of further shuffling could restore them to their ordered sequence. Only intelligent interference could do so. "Whenever anything happens which cannot be undone," he said, "it is always reducible to the introduction of a random element analogous to that introduced by shuffling."

As a further example he described a vessel partitioned into two halves, with air in one but none in the other. A valve is opened to permit the air to rush into the vacuum of the neighbor chamber. "For the moment all the molecules of air are in one half of the vessel," he wrote; "a fraction of a second later they are spread over the whole vessel and remain so ever afterwards. The molecules will not return to one half of the vessel; the spreading cannot be undone. . . ." Only some organized intrusion from outside can reverse the process. "This occurence," he continued, "can serve as a criterion to distinguish past and future time." If the process could have been recorded on film, a sane observer would have no difficulty deciding in which direction to run it. Anyone who thought it should show the air all rushing back into one chamber, leaving the other a vacuum, "had better consult a doctor," Sir Arthur said.

The introduction of randomness occurred not when gas first entered the empty chamber and raced toward the far side. It was then still "organized"—all molecules moving together. But then they hit the far wall and scattered. The organization was lost. That they should all, by some remarkable coincidence, end up in paths bringing them back into the first chamber was a possibility—but an almost infinitesimal one: "The adverse chance here mentioned is a preposterous number which (written in the usual decimal notation) would fill all the books in the world many times over."

By the same argument an egg that has been dropped on the floor could be reassembled. Each molecule of the yolk, white, and eggshell could be sent back along the route of the splatter to form an intact egg and have it fly back up into one's hand; but it is reasonable to suppose this would never happen. (An observation regarding black holes considered of considerable theoretical importance concerns their special relationship to entropy. A Princeton student, Jacob D. Bekenstein, while working on his doctoral thesis, noticed that the law specifying the inexorable way a black hole increases its area from the acquisition of new material, but can never decrease it, was strikingly similar to the rules governing entropy. Hawking then showed that the laws are, in fact, identical.)

A significant feature of entropy is its seeming dependence on some form of external influence. Its increase can to some extent to eliminated by isolation. Thus, as Eddington noted, a moving picture of the orbital motion of the earth around the sun could be run backward (showing it clockwise instead of counterclockwise) and would conform just as well to Newton's laws as if seen the other way. The earth's orbital motion (thank goodness) is largely isolated from outside interference. On the laboratory level, as pointed out by Thomas Gold at Cornell University, if an experiment is conducted in such isolation that no energy escapes, an increase in entropy is avoided once residual influences are dissipated. "Interference from outside," he says, clearly has something to do with it. "If we take any system and isolate it from all external influence completely and for a very long time and then take a series of snapshots, there will no longer be any way of deciding on the sequence from a subsequent examination of the pictures."

The implication, in the view of Gold, his old colleague Hermann Bondi, and others, is that the increase in entropy—and hence the arrow of time—are related to the expansion of the universe. As Gold put it, addressing a joint session of the American Association of Physics Teachers and the American Physical Society: "The essential property that the universe has is clearly that the sky is dark and that it will absorb radiation without limit," its darkness being a by-product of its expansion.

The debate on why the night sky is dark goes back to 1826, when the German astronomer Wilhelm Olbers stated his famous paradox. If, he said, there are an infinite number of stars scattered through the universe, there should be no

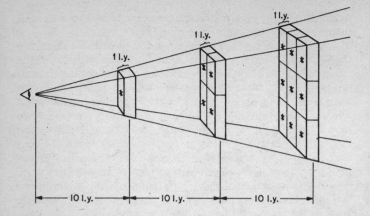

Olbers' paradox can be illustrated by a succession of boxes, ten, twenty, and thirty light-years distant. If the density of stars were uniform and there was one star in the box ten light-years away, there would be four stars in the four comparable boxes seen twenty light-years away and nine in the boxes visible at thirty light-years. In other words, the number of stars in the field of view would increase according to the square of the distance. While their individual brightness would fall off at the same rate, the two effects would cancel one another and the sky would be brilliant with an unbroken panoply of stellar radiation. *(California Institute of Technology)*

night; the sky would be bright even when the sun has set. While the observed brightness of stars falls off according to the square of their distance, the number of stars in any one patch of sky, Olbers reasoned, should, at the same rate, become more numerous with increasing distance, canceling the dimming effect.

Olbers attributed the darkness of the night sky to intervening dust, but (reapplying the problem to galaxies rather than stars) the discovery that the universe is expanding provided a more acceptable explanation. If dust were to blame, the effect presumably would be patchy; some parts of the sky would be brighter than others. But in an expanding universe light from the more distant galaxies becomes increasingly reddened and weakened by their motion away from us uniformly in all directions.

In the local environment, according to Gold's analysis, it is the nature of radiation that manifests the arrow of time: "Radiation in the world," he said, "is almost everywhere almost all the time violently expanding"—a phenomenon ultimately made possible by expansion of the universe.

If, then, the expansion is reversed, will time behave as in a moving picture run backward? Eddington noted in the 1920s that the steady increase in entropy represents a "running down" of the universe. Might a collapse be the "winding up" phase? Will entropy decrease and the arrow of time flip over? Will energy flow from cold bodies to hot ones? Will people stand in front of the fire to cool off? Will stars be heated, not by their internal fires, but from radiation "falling" inward as the universe collapses? Might the sense of time be reversed in such a way that, while to an external observer it seemed to run backward (people living from grave to cradle, so to speak), to those within the universe all would appear normal? Such concepts, as noted by Gold, seem "wildly improbable." But that does not answer the questions posed. As John Wheeler put it in 1975:

No one doubts that entropy increases, stars pour out energy, evolution moves foward in time, and memory contains only the past—and that all this development goes on while the universe is expanding. But the evidence is powerful that the expansion of the universe is slowing down and that there is truth in Einstein's views that this expansion will come to a halt and be followed by a phase of contraction.

As dynamic time marches forward, what will happen then to statistical and biological time? Will they continue to point in the same direction or will they point in opposite directions? In the one case, it will appear to be expanding, simply because a moving picture of contraction run backward looks like expansion. Many colleagues agree that the question is open and that the answer is one of the great puzzles of our day. . . .

John Archibald Wheeler lectures on black holes. *(Joseph Henry Laboratories, Princeton University)*

Paul Davies has pointed out that, according to general relativity, time travelers—he calls them "temponauts"—can voyage into the future through high velocity round-trip space travel. Yet it is difficult to prove in a rigorous way that the opposite is impossible. If, however, one could travel into the past, that would raise havoc with causality. A person could kill his parents before his own birth.

An intriguing suggestion has been made by Davies, namely that time reverses its arrow, not at maximum expansion of the universe, but when the latter has collapsed and erupts in rebirth. Thus the universe oscillates back and forth between phases when time runs forward through both expansion and collapse, then runs backward in a "subsequent" phase. The microwave glow that surrounds us, then, would be neither the residue of the last Big Bang, nor starlight that had leaked through from the previous incarnation of the universe, as proposed by Hoyle. It would be starlight that accumulated in the next universe and, flowing backward in time, had passed into this one! In a commentary on this proposal *Nature* termed it "more attractive" than the idea of time reversal at the ill-defined moment of maximum expansion. The instant of maximum collapse would be very sharply defined indeed, whereas when the expansion slowed to a stop and the clusters of galaxies began falling together again, the effect would become perceptible only by degrees.

In that case, as noted by Steven Weinberg, astronomers (if there are any) would first observe a shift toward the blue in the wavelengths of light from nearer clusters of galaxies. More distant clusters, being observed at an earlier stage in the history of the universe, would still be red-shifted. Initially, he wrote in his book, *The First Three Minutes,* there would be no significant increase in the radiation background. "However, when the universe has contracted to one hundredth its present size," he said, "the radiation background will begin to dominate the sky." The night sky will be as warm as the daytime sky (and, as contraction proceeds, black holes will probably proliferate). Then, after seventy million years, when the universe has contracted another tenfold, the expansion effect that resolved Olbers' paradox will be working in reverse, with devastating results. The sky will be so bright and hot that the molecules of air will break up into their constituent atoms and the latter will begin to shed their electrons. By then, or soon thereafter, life within this universe would have become impossible.

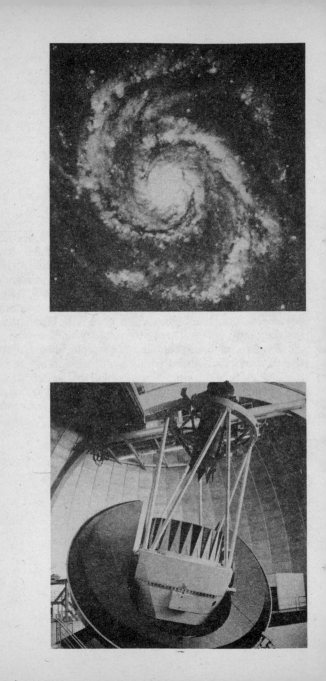

15

The "Stage of
the Universe Message"

The effort to determine by observation whether the universe
is open or closed, finite or infinite, destined for collapse or
endless expansion, dates back to the 1920s after the 2.5-meter
(100-inch) telescope on Mount Wilson, overlooking Pasadena
and Los Angeles, began its thirty-year reign as the world's
largest.

The challenge was to determine if the slowing of the expansion was sufficient to predestine an eventual halt and collapse. If one tracks the ascent of a soaring rocket, it is
possible to predict, from the extent to which its departure
from the earth is being slowed by gravity, whether it will fall
back or keep on going forever. The same test is applicable to
the universe.

One way to settle the matter would be to look far enough
out into space to see galaxies as they existed early in the life
of the universe and record from their motions the rate of expansion at that time. Comparison with the present rate would
then indicate the extent of slowdown. Another line of attack
would be to nail down as precisely as possible the present expansion rate and derive from that when the expansion would
have begun, assuming no slowdown at all. If, then, the actual
time of the Big Bang could be estimated—for example, from
the ages of the oldest stars or the time since the elements
(apart from primordial hydrogen and helium) were formed—
this, too, could be used to calculate the slowing. Only if the
time of the explosion was considerably more recent than indi-

cated by current expansion would any substantial slowing be indicated.

Allan R. Sandage of the Hale Observatories in California has spent much of his career seeking answers to these questions. His attempts to record from very distant (and therefore very ancient) galaxies the extent to which expansion was once more rapid was a sequel to the observations with the 2.5-meter telescope that enabled Edwin P. Hubble, beginning in 1929, to show a systematic relationship between the distances of the galaxies (based on their relative luminosities) and their rate of motion away from us (derived from the extent to which spectral lines in their light were shifted toward the red). When these two properties, observed from numerous galaxies near and far, were plotted diagrammatically, they formed a sloping line that indicated a steady increase in red shift with increasing distance—to be expected in the case of universal expansion—and the slope appeared to be uniform, implying a constant rate of expansion. Sandage and others, however, assumed that if the observations could be extended to a sufficient distance into space (and into the past), a change in slope should begin to show up, revealing a more rapid rate very long ago.

In search of this effect Sandage has used the brightest galaxies in clusters of galaxies, rather than galaxies picked at random. The largest assemblages of matter that we can see are not isolated galaxies but clusters of them and even "superclusters" formed of many clusters. Some clusters are themselves awesome in scope. The Virgo Cluster consists of about twenty-five hundred galaxies! Our own Milky Way Galaxy belongs to a group of about twenty that includes the Spiral Nebula in Andromeda (nearest twin of our own galaxy), the giant spiral known as Messier 33, and a number of dwarfs, including the two nearby Magellanic Clouds. The expansion of the universe seems to consist primarily of widening distances between clusters, rather than within the clusters themselves.

In his search for the fate of the universe Sandage hoped, by selecting only the brightest galaxy in each cluster, to depend on light sources that were as close as possible to uniform intrinsic brightness—what astronomers call a "standard candle." If at night one is trying to estimate relative distances

Allan Sandage. *(Hale Observatories)*

to many houses by the brightness of light bulbs in their windows, this can most accurately be done if it can be assumed that the brightest bulb visible in each house is of standard intrinsic brightness—say, a hundred-watt bulb. Their observed brightness, thus dimmed only by the distance, can then be used to estimate how far away each house is. When Sandage plotted the luminosities of the brightest galaxies in eighty-four clusters against their red shifts the points lay close to a straight line, delineating a steady increase of red shift with distance. A small degree of scatter either side of the line could be attributed, in part, to variations in intrinsic luminosity.

The problem then was to extend this line to a sufficiently great distance for the effect of faster early expansion to become evident. The extent of slowing from that early period can be expressed numerically (and is traditionally represented

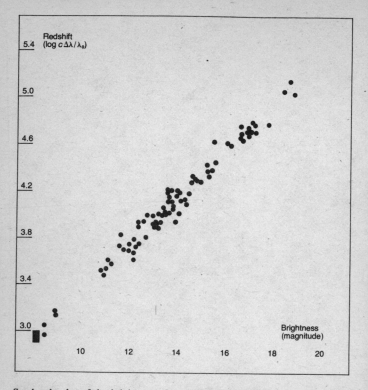

Sandage's plot of the brightness of very distant galaxies relative to their red shifts. As "standard candles" he used the brightest galaxy in each of eighty-four clusters of galaxies. Hubble's original data would lie within the small box in the lower left corner. *(Hale Observatories)*

by the symbol q_0). If q_0 is greater than .5, the slowing should be strong enough for a finite universe with closed (spherical) space-time curvature. If q_0 is less than .5, the universe is infinite and will expand forever along outward-curving (hyperbolic) lines. If it were exactly .5, space would be "flat" and the expansion would continue forever along straight lines. Finally, if q_0 were —1, the steady-state model would prevail.

When the quasars were discovered, it seemed that they

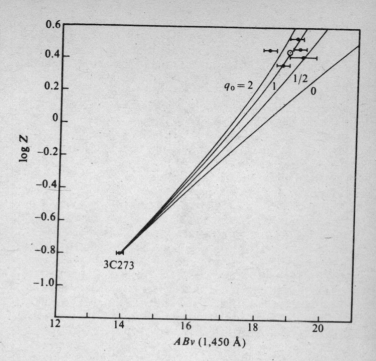

The extent to which recession velocities (red shifts) of distant objects become abnormally high at great distances (and look-back times) should indicate whether expansion has slowed enough since then to imply a closed, finite universe. The slowing is expressed by the term q_o. Any value of q_o less than $1/2$ indicates insufficient slowing to bring the expansion to a halt. Here red shifts (represented by the term z and expressed logarithmically) for five very distant quasars and a near one (3C 273) are plotted relative to their brightness in light originally emitted at the ultraviolet wavelength of 1,450 angstroms. The distant quasars all show red shifts greater than those in an open universe. As noted by those at The Johns Hopkins University who published this diagram in 1977, to conclude from it that the universe is closed would be "premature." It is shown here chiefly to illustrate how the q_o test is being applied in ongoing observations. (*A. F. Davidsen, G. F. Hartig, and W. G. Fastie in Nature*)

might at last provide the answer. If they were as distant as their red shifts implied, they should make it possible to extend the test for slowing back to a very early stage in the history of the universe. The Hale Observatories of Mount Wilson and Mount Palomar undertook an intensive search for more quasars, and this brought to light what Sandage referred to as "interlopers"—objects that resembled the quasars in other respects but were not observed at radio wavelengths. It has, in fact, now been found that some 95 per cent of quasars are radio quiet. Such objects, Sandage reported in 1965, seemed to extend the "look-back time" more than 90 per cent of the way to the Big Bang, reaching in distance (if the universe is closed) 63 per cent of the way to the "horizon"— the limit of observation at time zero.

Hopes for using the quasars to look for evidence of early, more rapid expansion were dimmed when the brightness of large numbers of them was plotted against their red shifts (presumably indicating their rate of recession). When this was done with ordinary galaxies, as previously noted, the relationship was strikingly uniform, but with the quasars there hardly seemed any relationship at all. The astronomical rebels who doubted that quasars are extremely far away seized upon this to question the use of their red shifts as distance indicators. Others noted that quasars often undergo radical changes in brightness, indicating that they may be too unstable to serve as "standard candles."

Sandage therefore concentrated on the brightest galaxies (giant ellipticals) in very distant clusters. By 1970 he saw what he took to be a hint of slowing—a q_o of 1.2 with an error margin of 0.4, implying a firmly closed universe. This, he reported, should not be taken "very seriously," but it was seized upon by those who wanted a closed, oscillating universe. In about twenty-five billion years, they said, collapse will begin, ending in a new Big Bang some sixty-five billion years hence (assuming the universe "bounced" instead of vanishing into a singularity).

The difficulty, as Sandage realized, was uncertainty regarding changes that may have taken place in the brightness of galaxies as they got older. Only if those sampled near and far had remained "standard candles" over the billions of years of their life histories would the luminosity-velocity relationship be meaningful (without some sort of reliable correction). He attempted such corrections, based on what was then known of the life histories of stars, and, with his colleagues

Jerome Kristian and William Westfall, continued to assemble data on distant galaxies. By 1977 they had compiled observations on 139 of them. Their luminosity-velocity relationship suggested a q_0 of 1.6 with an error margin of 0.4, but this time Sandage stated emphatically that no conclusion could be drawn from the data until the evolutionary effect was known.

At the Kitt Peak National Observatory in Arizona C. Roger Lynds has undertaken an ambitious attempt to measure this effect, using the observatory's four-meter telescope (exceeded in size only by the Hale Telescope on Palomar and the new Soviet six-meter instrument and matched by its sister at Cerro Tololo in Chile). His goal was to scan extremely distant (that is, ancient) galaxies and see if the brightness of individual points on their images was the same as with nearby galaxies. Distant light sources are dimmer than those nearby because (apart from a relativity effect for which a correction can be made) they occupy a smaller part of the field of view (according to the square of their distance). But points within their images should not be dimmed by distance. The scanning system used by Lynds included an image intensifier, a vidicon tube to convert the image to electric impulses, two computers to process the resulting signals, and television displays. The project is a continuing one.

Meanwhile, groups in the United States and China have reported finding ways to select certain classes of quasars that could be used as standard candles, thus extending the range of observation much farther into the past. They believe efforts to find a systematic relationship between distance and recession velocity for the quasars failed because the objects being observed were a mixture of many types, each with its own intrinsic luminosity.

One group, at The Johns Hopkins University, has taken advantage of a discovery by Jack A. Baldwin of the Lick Observatory that one class of quasars suitable as standard candles can be defined in terms of their brightness at certain ultraviolet wavelengths (such as the Lyman-alpha emission from hydrogen) relative to the background light of the quasar. There was evidence that those meeting this test were all of comparable intrinsic brightness. For the very distant quasars ultraviolet light, which normally cannot penetrate the atmosphere, is shifted toward the red sufficiently to be observable

The globular cluster Omega Centauri (NGC 5139), photographed by the four-meter telescope on Kitt Peak. *(Kitt Peak National Observatory)*

from the ground. This, however, is not true for the nearer quasars, and therefore a forty-centimeter telescope, the largest such instrument flown on a rocket up to that time, was launched from White Sands in April 1977 to look at the nearest quasar of all—3C 273. Using this one observation as the basis for comparison with the very distant quasars, a q_o of 1.0 was derived, but it was obviously very tentative. As the experimenters (Arthur F. Davidsen, George F. Hartig, and William Fastie) pointed out, pending further observations "it would be premature to conclude that the universe is closed." Such a quasar analysis has subsequently been done at the Lick Observatory, using eighty-four quasars near and far, with similar and somewhat more definitive results.

In China astronomers at the University of Science and Technology have reported on the use of another luminosity indicator. This is the extent to which radio components of quasars have become separated (assuming that their separation—perhaps following an explosive event—indicates the evolution of their luminosity). They used twenty-six quasars for which such data were available, grouping them according to the extent of separation and then using each group to look for evidence of retardation in expansion of the universe. They found that the intrinsic luminosity fell off systematically with increasing separation of radio components and that a q_o of 1.38 (meaning a closed universe) was indicated.

A basically different tactic in seeking to learn the fate of the universe is to determine the present expansion rate with precision and compare it with indicators of elapsed time since the Big Bang. If the current expansion, run backward to time zero, almost coincides with age-of-the-universe determinations, then clearly there has been little slowing.

Hubble's efforts to determine the current expansion rate—now known as the "Hubble constant"—have been continued by Sandage and others. From 1936 to 1952 the estimates, based on observations of hundreds of galaxies at a wide range of red shifts, implied a period of less than two billion years since the explosion—less time than age determinations for the earth. It was then found that the distance scale in use was systematically too short. The most recent calculations by Sandage and his colleague G. A. Tammann have galaxies 1,000 parsecs (3,260 light-years) apart separating at 50 kilometers a second. Run backward without allowance for slowing this puts the primordial explosion twenty billion years ago. (Some, notably the French-born astronomer Gerard de Vaucouleurs, believe the expansion is twice that fast and the age, without slowing, would be closer to ten billion years.)

If Sandage's estimate of twenty billion years, without slowdown, is correct, the actual elapsed time should be no more than thirteen billion years if there had been sufficient slowing to indicate a closed universe. David N. Schramm and his colleagues at the University of Chicago, using various assumptions regarding the ages of the oldest stars (notably those in globular clusters) and the time since elements heavier than helium were first formed, have placed the age of the universe somewhere between 13.5 and 15.5 billion years, implying that it is open.

Seemingly strong observational evidence for an open universe was presented at the 1974 Texas Symposium on Relativisitic Astrophysics. Like the first of those meetings, called eleven years earlier to ponder the discovery of quasars, it was held in Dallas. James E. Gunn of the Hale Observatories and the California Institute of Technology presented what has been described as the "State of the Universe Message" traditional at such meetings.

What seemed the most persuasive evidence for an open, infinite universe had come from measurement of the deuterium mingled with hydrogen in gas clouds of the Milky Way Galaxy. The relative abundance of deuterium, a heavy form of hydrogen, was regarded as a strong clue to the density—and hence destiny—of the universe right after the Big Bang. Hydrogen, the simplest of all atoms, normally consists of a single proton with its attendant electron, but sometimes the proton is paired with a neutron, which does not alter the chemical properties of the atom but makes it twice as heavy. That is deuterium. The next heavier atom is helium, whose nucleus is formed of two protons usually mated with one or two neutrons. Deuterium and then helium are believed to have been synthesized by high-energy collisions during the early minutes after the primordial explosion. However, the link between proton and neutron in deuterium is so weak that in the intense heat of the first three minutes those particles could not remain joined and, according to present reconstructions of what happened, no deuterium survived. Then, between three and four minutes after the bang, it cooled enough so that protons and neutrons stuck together, forming deuterium. Much of the deuterium almost immediately acquired added protons and neutrons, becoming helium. The amount of deuterium that escaped this transmutation would have depended closely on the density of nuclear particles in the expanding cloud. If the density was great, collisions with protons would have converted almost all the deuterium into helium. Seeing how much deuterium is around now, it was thought, would indicate whether or not the universe, right after its birth, was dense enough to remain closed.

The measurement that particularly impressed those at Dallas had been made with Princeton University's 81-centimeter telescope riding the Copernicus satellite in earth orbit. Instrumented for observations at ultraviolet wavelengths that do not penetrate the atmosphere, the telescope had been aimed at Beta Centauri, the tenth brightest star in the sky and, among those ten, the hottest. The purpose was to record the relative extent to which hydrogen and deuterium along the 264-light-year path between that star and the earth had absorbed its light at their characteristic wavelengths, thus indicating their relative abundances. As with observations being made by radio astronomers from the earth it was found that the amount of deuterium was surprisingly high—roughly 1.4 atoms for every 100,000 of hydrogen. This, the Princeton group reported, indicates a primordial density that "falls a factor 27 short of the critical density for closing the Universe." The inferred present average density would be the equivalent of only one hydrogen atom per ten cubic meters.

At the Dallas meeting Hubert Reeves of the French nuclear research center at Saclay, near Paris, said: "Anybody who wants to believe in the Big Bang producing a closed universe had better invent a way of making deuterium, and that's not easy, I can tell you." Stirling Colgate, the specialist in supernova explosions, had proposed that they could do the job, but Reeves dismissed this as improbable. On the contrary, it was argued that deuterium had probably been destroyed as the galaxies evolved, so that there was even more to begin with than now observed.

On the eve of the Dallas meeting at which Gunn summarized "the state of the universe" he and several other participants had published their arguments for an open universe, stating that the only way to escape the conclusion that the density of the universe was low when the deuterium was formed would be the existence of matter "hidden in a collapsed form from a very early epoch indeed"—that is, a multitude of black holes.

Adding to the difficulties of those holding out for closure was the present "temperature" of the universe as indicated by the microwave glow seemingly left from the Big Bang. The refined estimate was now 2.7 degrees above absolute zero, whereas Dicke's group at Princeton, based on Peebles' calculation of primordial helium production, had expected that in a closed, oscillating universe it must now be at least ten degrees.

The troublesome question regarding the extent to which galaxies may have become dimmer with age was dealt with at Dallas by Beatrice M. Tinsley, a British-born astronomer then at the University of Texas and a coauthor of the newly published case for openness. She had concluded that galaxies had been sufficiently brighter in their youth to make the slowdown seem much greater than it really was.

As noted earlier, efforts to determine q_0 (the extent of slowing) consist essentially of comparing the luminosity of very distant galaxies (as an indicator of their distance) with their velocity of recession (based on their red shifts). If, at very large apparent distances, they are receding considerably faster than would be the case in a universe whose expansion rate has remained constant, this would imply that subsequent slowing has been sufficient to close the universe. But if the luminosity of galaxies long ago was greater than now, then those seen far away (and hence far into the past) were brighter, and therefore more distant, than assumed. The evidence for substantial slowdown would vanish.

Tinsley argued that, when the galaxies were still packed with bright young stars, they were probably so much more luminous than today that very little evidence for slowing is apparent. The implication, she said, was an open universe by an "embarrassingly" large margin. (In fact, the effect seemed so strong that she and others thought, for a time, that the expansion had speeded up instead of slowing, leading to a revival of Einstein's concept of a cosmological term that was pushing the galaxies apart.)

Probably the oldest problem, for those seeking to close the universe, was finding by direct observation at least a substantial fraction of the material needed to do so. The gravitational attraction fighting against eternal expansion (as spelled out in the 1920s by Aleksandr Friedmann) is determined by the average density of the universe—the number of grams or tons of material per cubic light-year. Considering the vast reaches of seemingly empty space between galaxies—and even within galaxies themselves—the average density seemed to fall short by a factor of fifty.

There was, however, a hint of substantial amounts of hidden material, at least within the clusters of galaxies. Otherwise the clusters should not exist. When the motions of galaxies within a cluster are analyzed, they are found to be moving like a swarm of angry bees. If they are kept from flying off in all directions by nothing more than the gravity of their stars and other observed material, they should have dispersed long ago. The observed material is only one twentieth of that needed to hold the clusters together. Jeremiah F. Ostriker of Princeton proposed to the Dallas meeting that each galaxy is enveloped in a giant, invisible halo of stars and other material that multiply the mass of the galaxy tenfold. Such stars could be so small and dim that they would never have been observed. As stated by Tinsley, Gunn, and the other authors of the newly published case for an open universe: "Possibly most of the mass of the universe resides in this silent majority of small stars." Nevertheless, they pointed out, this added material fell far short of what was needed: "Galaxies themselves cannot close the universe."

Summing up in his "State of the Universe Message," Gunn said that, while plausible rebuttals can be made to some of the open-universe arguments, failure of all of them "requires a rather bizarre combination of conditions, which seems unlikely at best." He and his colleagues, mindful of the deep-seated reluctance to accept an open universe, prefaced their published paper with a quotation from Lucretius, the great natural philosopher of the last century before Christ:

Desist from thrusting out reasoning from your mind because of its disconcerting novelty. Weigh it, rather, with a discerning judgment. Then, if it seems to you true, give in. If it is false, gird yourself to oppose it. For the mind wants to discover by reasoning what exists in the infinity of space that lies out there, beyond the ramparts of this world. . . . Here, then, is my first point. In all dimensions alike, on this side or that, upward or downward through the universe, there is no end.

They concluded by saying cautiously: "The objections to closed universe are formidable but not fatal; a clear verdict is unfortunately not yet in, but the mood of the jury is perhaps becoming perceptible."

After the evidence had been presented at Dallas a group of young astronomers and I went to dinner, and I passed to each a slip of paper asking them to indicate their own verdicts by writing on it "open" or "closed." When the ballots were counted the vote was overwhelmingly "closed." As John Faulkner put it in his report on the meeting to *Nature,* "Perhaps cosmologists should always state, 'These are the opinions upon which I shall base my facts.' "

By the late 1970s, there was a perceptible pendulum swing back from the open-universe concept. As Beatrice Tinsley (after joining the Yale faculty) herself wrote in *Physics Today,* a few years earlier there seemed to be "consistent, albeit tentative evidence for an open, ever-expanding universe. Since then, further data and theory have inevitably conspired to blur the appealing simplicity of that picture." Among the new developments was the argument by Ostriker and Scott Tremaine that as the central galaxies in clusters grow older they "swallow" their neighbors and become brighter. This could cancel—or even reverse—the loss in brilliance that she had believed undermined the argument for an observed slowing, although she still leaned toward an open universe.

The recent efforts in China and the United States to use quasars as "standard candles" have kept alive the possibility that the universe is, in fact, closed. On the other hand deuterium measurements by Arno Penzias, codiscoverer of the Big

Bang residual "glow," using the new seven-meter radio telescope of Bell Laboratories, have strengthened the deuterium argument for an open universe. New discoveries, however, in high-energy physics experiments (the possible existence of a "heavy lepton" and its associated neutrino) have suggested to some theorists that present reconstructions of what happened in those critical early minutes—the foundation of the deuterium argument—may have to be revised. Thus the matter remains unresolved.

"To some people," wrote Beatrice Tinsley after summing up the evidence for expansion in her *Physics Today* analysis, "the prospect of a monotonically expanding universe is philosophically bleak; to others, the thought that a definitive model for the universe itself could be reached is incredible (if not absurd) and to many the empirical basis of the foregoing conclusions is altogether insecure."

Not everyone abhors the prospect of perpetual expansion. A statement on behalf of those prepared to welcome such a fate was made a half century ago by Sir Arthur Eddington during a series of lectures (later published as *The Nature of the Physical World*):

> I am no Phoenix worshipper [he said]. This is a topic on which science is silent, and all that one can say is prejudice. But since prejudice in favour of a never-ending cycle of rebirth of matter and worlds is often vocal, I may perhaps give voice to the opposite prejudice. I would feel more content that the universe should accomplish some great scheme of evolution and, having achieved whatever may be achieved, lapse back into chaotic changelessness, than that its purpose should be banalised by continual repetition. I am an Evolutionist, not a Multiplicationist. It seems rather stupid to keep doing the same thing over and over again.

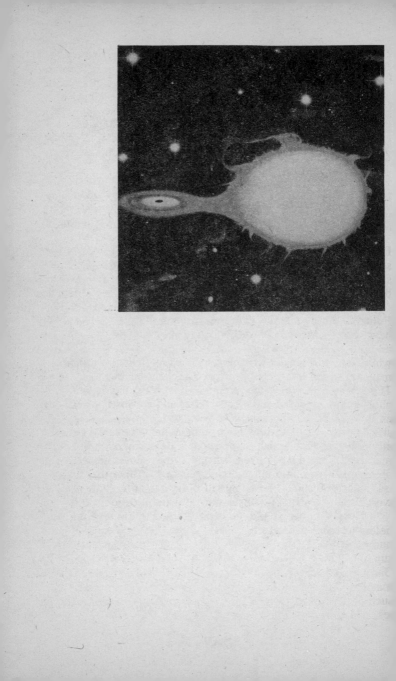

16
Black Holes—
Everywhere or Nowhere

During the past year or two there have been so many reports suggesting the presence of black holes, near and far, that, despite evidence for an eternally expanding universe, the possibility remains that it is held together by such objects.

Among the more sensational proposals has been one, by Edward Harrison of the University of Massachusetts in Amherst, that the sun and its planets may very slowly be circling an unseen companion—possibly a black hole. He noted that while other pulsars are slowing their rates in a systematic manner, those in one quarter of the sky are not doing so. This, he suggested, could be because the solar system, including the earth, is moving toward them in its orbit around an unseen object—a Doppler effect. It might, he said, be a star of extremely low luminosity, or a black hole. "I personally find it hard to believe that a star so close can exist and remain undiscovered," he wrote. "On the other hand pulsar observations of extraordinary precision imply that it might exist, and therefore a search for a companion star is perhaps worth undertaking."

There have been a variety of proposals as to how black holes, in some cases of enormous mass, could exist in great numbers. The Soviet theorists Zeldovich and Novikov have suggested that miniholes or "seed" black holes in the Big Bang fireball could have "rapaciously" swallowed up nearby material, becoming as massive as a million billion suns. If such objects, now scattered through space between the galaxies, accounted for all the missing mass, they would, on the average, be more than thirty thousand light-years apart, and unless, like the X-ray binaries, they were still swallowing great gobs of matter, they would be difficult to observe.

One way to detect supermassive black holes in splendid isolation, far removed from galaxies and clusters of galaxies, was proposed in 1973 by Gunn and William Press, then at Cal Tech. This was a focusing effect to be expected from the gravitational fields of such objects. Light diverging from a distant quasar or galaxy would be bent as it passed a black hole, focusing it toward the observer in the manner of a lens. It was improbable that a giant black hole would be exactly centered on the line-of-sight to a distant light source, whereas if it were close enough to that line, the focusing should form a second image close to, but not directly superimposed on, one that had traveled a direct route, clear of the black hole's influence. The effect should apply both to optical and radio sources. While the separations would depend on the mass of the object responsible for the focusing, they would be very narrow—comparable to one ten-thousandth of a second of arc. Nevertheless, Press and Gunn said, this should be observed with very widely separated radio telescopes (the method known as very-long-baseline interferometry or VLBI). To date no systematic search for such double images has been carried out.

Another optical effect cited by Tinsley for a tightly closed universe might be observed if one could look far enough to see the "antipole" of the curvature. In such a case the whole universe would cause the focusing. Light emitted from a distant object, such as a quasar, would arrive via a multitude of routes, as does sound traveling around the inside of a sphere from a source directly opposite (an effect I experienced, to my astonishment, in the deep submersible *Alvin*, when, within the spherical pressure vessel, the voice of my companion seemed extraordinarily amplified). As she pointed out, the abnormally bright objects that, in such a situation, should be seen at maximum red shifts are not observed. This does not rule out a closed universe, but indicates that we apparently cannot see as far as the antipole.

For many years astronomers have speculated that much of the "missing mass" might consist of invisible material between the galaxies. One proposal, by P. J. E. Peebles of

Princeton, was that it might be in the form of "unborn" galaxies—clouds of dust and gas that had not yet condensed sufficiently to produce stars—or, at the opposite extreme, "dead" galaxies whose stars had run through their life cycles and become too dim to be observed at great distances. That such galaxies might lie between us and very distant quasars, he said in 1968, could explain why light from those quasars had, in some cases, been absorbed by gas moving away from us considerably less rapidly than the quasar itself.

A more traditional concept has been of dust and gas spread thinly and uniformly through the apparent void between galaxies (or clusters of galaxies). Efforts have been made to observe whether such gas robs the light from distant galaxies of wavelengths that would be absorbed by the gas, or whether dust particles and gas molecules scatter the shorter (bluer) wavelengths—the phenomenon that makes the sky blue. This would cause the galaxies to appear brighter in red light than in blue (a phenomenon unrelated to the red shift). No such universal absorption or light scattering has been observed, but, as noted earlier, when systematic X-ray scanning of the sky became possible, starting with *Uhuru,* a universal background glow was observed (separate from the residual Big Bang glow, which is at much longer wavelengths). While this X-ray background is diffuse, its total energy is considerable, being a thousand times greater than that of radio waves reaching us from all the quasars and other distant sources.

What "cosmic inferno," asked Herbert Friedman, could possibly have been responsible for this glow? By 1977 some on the *Uhuru* team in Cambridge, Massachusetts, as well as those at the University of Leicester making measurements with Britain's Ariel 5, suspected that a substantial part of the glow originated long ago in countless Seyfert galaxies. The latter, as noted earlier, have brilliant nuclei in which violent activity occurs—or did, when they and the universe were young. (They may, some believe, be the source of a faint gamma-ray glow that pervades the heavens.) Harvey Tannanbaum of the *Uhuru* group proposed that each Seyfert nucleus contains a black hole of from one million to ten million solar masses whose gobbling up of nearby gas could account for the X-ray emissions.

The *Uhuru* and Ariel 5 observations also seemed to

show that strong X rays were coming from very hot gas spread through the more densely populated clusters of galaxies. It was noted by two of Sir Martin Ryle's colleagues at Cambridge (S. F. Gull and K. J. E. Northover) that the mass of this gas "might well be a substantial fraction of that of the whole cluster." Hot gas, they continued, "may therefore be a major constituent of the universe, so that independent confirmation of its existence is extremely important."

One way to seek confirmation would be to find out whether the hot gas presumed to lie within the clusters had altered the microwave glow from the Big Bang. When a gas is extremely hot—at many millions of degrees—its electrons move so fast and furiously that if they hit a photon (such as a light wave or a radio wave) the latter can emerge from the encounter with more energy than it had to begin with. The effect (known as inverse Compton scattering) can be likened to the added momentum sometimes acquired by a lightweight ice-hockey player in a glancing collision with a beefy one. In the case of the residual Big Bang radiation there would be a redistribution of wavelengths in the glow coming from the direction of such clusters, more of the waves appearing at higher energies than from other parts of the sky. Observations with a radio telescope of the Chilbolton Observatory, Gull and Northover reported, seemed to show this effect and thus "strongly suggest the presence of hot gas filling clusters of galaxies at a temperature of about a hundred million degrees.

A more extensive distribution of hot gas was deduced in 1977 when the *Uhuru* group, still led by Riccardo Giacconi and now installed at the Harvard-Smithsonian Center for Astrophysics, issued its *Fourth* Uhuru *Catalogue of X-ray Sources,* compiled from the flood of data acquired before *Uhuru* was shut down by radio command in 1974. It listed 339 new sources, including 53 that appeared to be distant clusters of galaxies (many, as already noted, grouped into "superclusters"). Giacconi and his colleagues, from their analysis, concluded that in three superclusters (and probably all of them) hot gas envelops the whole assemblage. This implied the presence of gas with five to ten times more mass than that of all the galaxies themselves—enough to hold the supercluster together and contribute at least somewhat to the mass needed for closure of the universe.

Even stronger evidence has come from an experiment in which the Navy group, with its instruments aboard HEAO-1, recorded X-ray emissions as the moon passed across an area of the sky occupied by two clusters of galaxies in the constellation Aries, the Ram (Abell 401 and Abell 399). The cutting off of X rays by the moon showed not only that they originate in and around both galaxies but that the emissions are strongest from the region midway between them implying extensive distribution of the hot gas. "The new X-ray evidence," said Friedman, "places most of the 'missing mass' in primordial gas left over from the epoch of galaxy formation."

One proposed explanation for the diffuse X-ray glow is that a substantial part of it originates from extremely high-energy collisions in a very thin but very hot gas that is everywhere—not just associated with clusters or superclusters. Such collisions generate X rays by the process known as *bremsstrahlung* ("braking radiation"). This, rather than black holes in countless Seyfert galaxies, was the explanation favored by George Field, head of the Harvard-Smithsonian Center for Astrophysics. The gas, he believed, had been heated by explosions and other violent processes (including, perhaps, the collapsing of numerous black holes) that seem to have been frequent within galaxies when the universe was young. It is suspected for example, that quasars, with their highly variable luminosity and their tendency to be most common at very great distances (and long look-back times), manifest galaxies in the wild throes of adolescence. The twin sources of intense radio emission apparently ejected in opposite directions and at high velocity from the radio galaxies were further evidence that vast amounts of energy have flowed into space.

Field and his colleague, Stephen C. Perrenod, have tried to estimate how much gas in the vast reaches between clusters has been heated in this manner. Is it sufficient to hold the universe together? If the amount of energy that stirred up the gas in the first place were known, and therefore its present temperature, it would be possible to assess the density required to produce the observed X rays. Using reasonable estimates of the original energy input, they put the temperature at from a hundred million to a billion degrees and concluded from this that the most likely density was about half that required for closure. However, they said, considering the uncertainties that remain, the amount still "could be" enough to prevent eternal expansion.

As in the earlier British analysis, Field and Perrenod looked for the effect that a hot gas would have on the residual Big Bang glow—but this time it would be evident in all directions, not just toward galactic clusters. Of critical importance were measurements made from balloons high enough in the atmosphere to record the very shortwave (submillimeter) component of the glow. The balloon observations available seemed to show the effect. Observations with HEAO-2, however, indicate that distant quasars make a substantial contribution to the X-ray glow.

A continuing debate regarding the existence of supermassive black holes concerns the possibility that they form the cores of at least some globular clusters. The latter are almost perfectly spherical swarms of stars—often several hundred thousand to a million or more—and represent a considerable constituent of the galaxies. There are, for example, more than 130 of them distributed through a vast spherical region centered on the nucleus of our own Milky Way. Their stars are very old—of the first-generation type deficient in heavy elements, not having inherited those synthesized in other stars or in supernovas.

In the 1960s it was noticed that the cores of the most concentrated globular clusters are remarkably bright, as though there is an excess of mass there. Arne A. Wyller, then at the Bartol Research Foundation in Swarthmore, Pennsylvania, proposed that these are concentrations of stars held there by one or more black holes remaining from formation of the cluster. Astronomers at Princeton University then proposed that stars in the central region of a cluster would be so crowded that they would collide, in this manner forming a supermassive black hole representing perhaps 10 per cent of the total mass of the cluster. Finally, two Princeton neighbors—John Bahcall at the Institute for Advanced Study and Jeremiah Ostriker at the university—suggested that these black holes, with masses exceeding those of a thousand suns, would account for the X-ray emissions newly detected from five such clusters (a similar proposal was advanced by Jonathan Arons and Joseph Silk at the University of California in Berkeley).

The first X-ray observations of such objects had been made by *Uhuru*, but now a new generation of satellites with X-ray detectors was in orbit: Britain's Ariel 5, Orbiting Solar Observatory 7, the Astronomical Netherlands Satellite, and Small Astronomical Satellite 3, launched from the platform off Kenya. Not only did their observations point to several globular clusters as X-ray sources, but they also began to detect extremely short X-ray bursts from certain points in the sky—most of them in the general direction of the galactic core. These "bursters" quickly evolved into the most talked-about and perplexing features of the X-ray sky.

The first observation was made in 1971 with the Soviet satellite Cosmos 428. It was described in a paper published four years later in the Soviet literature and, apparently, overlooked by all researchers outside of the Soviet Union. In November 1975, Jonathan E. Grindlay of the Harvard-Smithsonian Center for Astrophysics was studying recordings made two months earlier by American and Dutch detectors on the Astronomical Netherlands Satellite when, to his astonishment, he found that the intensity of X rays from a source under observation had increased thirtyfold in a half second! Over the next ten seconds it then smoothly returned to the original level. A re-examination of recordings from the same source showed a similar burst eight hours earlier. The source seemed to lie squarely in the center of the globular cluster NGC 6624. (NGC refers to the *New General Catalogue of Nebulae and Clusters of Stars*, published in 1888 by J. L. E. Dryer, head of the Armagh Observatory in Ireland.) Another cluster, NGC 1851, also seemed to harbor a burster. By early 1979 thirty bursters had been found, all following a typical pattern of extremely rapid rise to full intensity (in less than a second), then smooth decay within a few tens of seconds. In that brief period a burster may emit as much energy, within a narrow X-ray band, as does the sun in a month at all wavelengths. While the bursts usually recur every few hours, an exception has been what its discoverers at the Massachusetts Institute of Technology refer to as "the rapid burster." In a form of celestial cannonading to date seen nowhere else in the sky it may fire five thousand times a day. The most powerful bursts are followed by a relatively long intermission, whereas weaker ones come in quick succession, as though trying to make up for their weakness—a process typical of situations where some steady input of energy or pressure is being relieved irregularly (as where a dripping faucet pauses longest when an unusually large drop has just fallen).

William Liller of Harvard aimed the new four-meter telescope of the Inter-American Observatory at Cerro Tololo in the Chilean Andes toward the rapid burster. Because it lay in the direction of the galactic center, where clouds of dust and gas intervene, he made a photographic exposure at infrared wavelengths (which more readily pass through such clouds) and recorded what he took to be a previously unknown globular cluster. This association of the burster with that globular cluster was further strengthened by HEAO-1 (High Energy Astronomy Observatory 1) after it entered orbit in 1977.

Although the sky is peppered with such clusters, no more of them have been linked to bursters, and it seems likely that many of the latter are related to some other kind of object that tends to reside closer to the central region of the galaxy than the globular clusters. Nevertheless, it is widely suspected that the bursters manifest gobs of material falling at extreme velocity onto collapsed objects, such as neutron stars or black holes. The extraordinary behavior of the bursters is reminiscent of Cyg X-1—the original black-hole candidate—whose intensity has been observed to double in a twentieth of a second.

When several theorists saw black holes as the most likely explanation for the bursters, Kenneth Brecher of the MIT dissenters struck back. The central object in each case, he said, need be no more than a few solar masses and, he added, "one must turn from black holes to some other, more physical and realistic model." It seemed to him that black holes were being used to explain every puzzle in astrophysics, including failure to observe the neutrinos, or "ghost particles," that, according to theory, should be generated by fusion reactions within the sun and then fly out, unimpeded, in all directions. (Perhaps, it had been suggested, they were being kept from escaping by an accumulation of miniholes that had fallen into the sun and coalesced in its core.)

"Despite the fact that in the past few years," Brecher wrote in *Nature,* "black holes have been offered up as a panacea for explaining everything from the Tunguska event, to the absence of solar neutrinos, to the energy emitted by quasars, and even as a solution to the terrestrial energy crisis, there seems to be little direct observational support for the assertion that X-ray bursters provide evidence for the existence of massive black holes."

At the 1978 "Texas" symposium in Munich Walter H. G. Lewin and Paul Joss of MIT proposed a "helium bomb" explanation for the bursters. As gas from a companion star falls onto a collapsed object, such as a neutron star, according to their hypothesis, it forms layers including one predominantly of helium. Periodically the helium becomes compressed and hot enough to fuse explosively, as does hydrogen in a bomb, but on a catastrophic scale, causing the X-ray flashes.

Evidence implying that at least some bursters contain black holes has been found by Friedman and his colleagues at the Naval Research Laboratory, using the massive HEAO-1 satellite. They have recorded trains of rhythmic X-ray flashes such as those predicted for blobs of super-hot gas spiraling in toward a hole. The source is the burster known as 1728–34 (for its celestial co-ordinates). It is not in a globular cluster. The flashing rate is twelve thousandths of a second, which Friedman believes is the time it would take a hot spot to circle a hole of twenty-five solar masses in its final spirals before vanishing into the hole. There is also evidence of a second, similar flasher.

Such black holes, however, are "small potatoes" compared to those that, it is now widely suspected, may be the powerhouses in the cores of galaxies, particularly those pouring out catastrophic amounts of energy. Some astronomers believe they are millions and even billions of times more massive than the sun.

It will be recalled that as early as 1963 (before black holes came into their own) Fred Hoyle and William Fowler proposed that the energy output of radio galaxies could arise from gravitational collapse in their cores involving from one hundred thousand to one hundred million solar masses. In 1970, at a "Study Week on Nuclei of Galaxies" held at the Vatican Observatory, Donald Lynden-Bell of Britain's Royal Greenwich Observatory (then on leave at Cal Tech) suggested that, in the early stages of a galaxy's evolution, friction between the fast-spinning interior part of the system and the slower-moving outer part would transfer angular momentum outward from the inner zone. This would slow the material of the inner zone which would spiral toward the center of the system, becoming increasingly dense and hot (a process, similar to that postulated for the disk around a "conventional" black hole, further elaborated by Bardeen of the University of Washington to take into account the effects of rapid rotation).

Quasars, according to Lynden-Bell, would be galaxies whose inner material has become dense and hot enough to shine with extraordinary brillance before vanishing inside their Schwarzschild radius. A third of the mass of the infalling material, he said, could be converted into energy in this manner. Once the supermassive black hole was formed it would continue to swallow material, lumpiness and friction in the surrounding disk sometimes causing it to "overeat." This would generate a sufficient burst of radiation to cause an explosion of the type seen intermittently in the nuclei of Seyfert galaxies (and especially in quasars like 3C 279, which flare up to many times their normal brilliance). Furthermore, Lynden-Bell added, "it is possible that violent overfeeding of a black hole" leads to an explosion of hot gas out along the rotation axis of the galaxy. Such gas streams could account for the odd structure observed when radio galaxies—those that are the most powerful sources of radio waves—are mapped by radio telescope. The strongest emissions, it has been found, do not originate in the galaxy itself but from regions far out on opposite sides of it along its rotation axis, as though from material blown out in opposing directions millions of years ago and at close to the speed of light. A classic example is the radio galaxy Centaurus A. Such speculations, however, said Lynden-Bell, "need stronger derivations before they can be applied usefully to radio outbursts."

That ejections of material from the cores of galaxies can

be continuous over prolonged periods, rather than explosively episodic (thus requiring almost unbelievable releases of energy), was indicated after mapping of the galaxy NGC 6251 by British and American radio astronomers. In a pattern typical of radio galaxies there are two regions of very powerful radio emission far out on opposite sides of the visible galaxy. Mapping by the radio astronomers at Cambridge, England, showed a long, narrow jet extending from the galaxy to one of those outlying radio sources (the latter actually is split in two). Then on July 23–24, 1977, three widely separated radio telescopes—at the Haystack Observatory in Westford, Massachusetts, the National Radio Astronomy Observatory in West Virginia, and Cal Tech's Owens Valley Radio Observatory in California—were aimed at the visible galaxy. Their recordings were timed by atomic clocks so they could later be combined to produce a detailed map of the region (the method serving as the basis of very-long-baseline interferometry).

This showed that the long jet mapped by the British originated in a tiny source at the very center of the galaxy. A "cosmic blowtorch" had thus been identified that extended more than 750,000 light-years from this central energy source to the distant blobs of radio-emitting material. Such a length indicated that the core has been pouring out its torrents of energy for considerably longer than 750,000 years.

While there remain a number of dissenters from the black-hole explanation for such energy production, there seems almost universal agreement that a single process of some sort accounts for all the wild behavior being observed within galaxies, including the quasars, their apparent cousins, the BL Lacertae objects, the tumultuous Seyfert galaxies, and the radio galaxies like NGC 6251. (Some suspect the same, on a far smaller scale, may apply to the core of our own galaxy.) Geoffrey Burbidge (who doubts the black-hole explanation) calls the energy source "The Machine." James Gunn of Cal Tech refers to it as "The Monster" and Martin Rees of Cambridge simply calls it "The Prime Mover."

A tantalizing clue to the nature of The Machine was reported to the 1978 "Texas" symposium (held in Munich). Roger Angel of the Steward Observatory said he had found that the polarization of light from some (if not all) BL Lacertae objects rotates in a systematic manner. In polarized light the waves all vibrate in the same direction. Under the influence of a magnetic field, the plane of polarization can rotate. As much as 30 per cent of the light from some BL Lacertae objects is polarized, Angel reported, and steady rotation of the polarization can be followed throughout a night of observing. While this has been seen in several such objects, the strongest evidence is from the BL Lacertae known as $1308 + 326$—one that is so distant it even lies beyond some quasars. The rotation rate is typically a few degrees per hour, as though something inside The Machine was itself rotating. The polarization may reverse the direction of its rotation from night to night, possibly because different parts of the system are being observed. A similar effect at radio wavelengths has been recorded by University of Michigan astronomers in the emissions from BL Lacertae $0235 + 165$, but the rate is much slower, requiring about one hundred days for a complete revolution.

The evidence for rotation is the first hint of an ordered process at work. It may be a key feature of The Machine that is concealed in galaxies and quasars. Theories for the energy production fall into two categories: "accretion" models and "spinar" models. The former envision gas falling into the gravity field of a supermassive object, such as a black hole, and being greatly heated. The spinar model is of a massive, spinning, magnetized object whose rotating magnetic field generates electromagnetic radiation. Rotation would be intrinsic to both systems.

When Rees, at the previous Texas symposium, summed up current efforts to explain the quasars, he focused on the black-hole hypothesis, proposing that such holes, with more than ten million times the mass of the sun, may be fed by the falling in of "entire stars." It was noted that black holes of very great mass—say, one hundred million times that of the sun—would not necessarily be very dense from the viewpoint of an external observer. Since, from that perspective, time within the hole would have virtually halted, the density there would be frozen at a low level—perhaps no greater than that of water. What a contrast to the neutron star, whose density may be millions of tons per cubic centimeter! To someone inside the hole, plunging toward a singularity, however, the density would rapidly approach infinity.

Rees and his colleagues have proposed that the energy source in Centaurus A, a relatively nearby radio galaxy, is a black hole of ten million solar masses (a less exotic explanation has been offered by a group at NASA's Goddard Space Flight Center). The radio emissions of Centaurus A vary within periods as short as a day, and the X-ray output changes markedly within a few days, implying a rather compact source. One possibility, they suggested, is that as a star comes close to a central hole it is broken up by tidal stresses (much as an astronaut, approaching such an object, would be torn apart). The material of the star, circling and finally falling into the hole, would generate the observed emissions. The radius of this core region would be only six hundred times that of the earth's orbit—relatively a speck on the scale of a galaxy. "The variability in the X-ray source," their report said, "may arise from changes in the mass flow rate, or amount and distribution of angular momentum in the infalling gas." They noted that, compared to such extremely powerful radio-emitting galaxies as Cygnus A, Centaurus A is a weakling. But, while emissions from the galaxy itself are relatively feeble, those from the satellite regions to either side of it (presumably ejected at an earlier stage) are a billion billion times stronger. Rees and his colleagues suggested that Centaurus A, with a black hole as its heart, may represent a typical late stage in the evolution of more powerful sources such as Cygnus A.

The strongest candidate so far for a supermassive black hole—one five billion times more massive than the sun—has been identified in the core of the galaxy Messier 87. (Known also as M 87, it is No. 87 among the 109 celestial features catalogued, beginning in 1760, by the French astronomer Charles Messier.) It is a giant elliptical galaxy whose mass is so great it seems to play a major role in holding together the 130 other galaxies of a cluster in the constellation Virgo. Despite its distance (sixty-five million light-years), its radio and X-ray emissions reach the earth with considerable intensity. Its core has been the suspected seat of violent activity because photographic exposures highlighting that brilliant region reveal jets emerging from it, one of them five thousand light-years long. (Rees thinks M 87 may be a former quasar tempered by old age.)

To explore what is at the core of M 87 an American-British-Canadian team undertook a project whose results were published in 1978. They used two of the large telescopes on Mount Palomar to map light intensity in the core region and the four-meter reflector on Kitt Peak to record the motions of stars in the core's vicinity. At such a distance individual stars cannot be identified, but the rapidity of their motions can be determined by the extent to which collective spectral lines from those stars are broadened by Doppler shifts. The faster their movements, toward and away from the earth, the broader will be each spectral line.

The observations could be made in great detail and with unprecedented precision because of a variety of new devices. (For the light intensity mapping these included silicon intensified target tubes and charge coupled devices; for the spectral observations an image photon counting system provided by University College, London, was used.)

Scanning by the group on Palomar revealed a sharp spike of light intensity at the center of the core. Those on Kitt Peak found that as their scanning approached the core there was a sudden increase in the velocity of star motions (from 278 to 350 kilometers per second), indicating that stars in the inner region are moving under the control of an enormously massive, superdense object. Such an effect was not seen when, for comparison purposes, they scanned the less dynamic galaxy NGC 3379.

They concluded that the observations "are entirely consistent" with the presence of a central black hole five billion times more massive than the sun and less than 100 parsecs (326 light-years) in radius. Applying this to their brightness measurements the Palomar group concluded that the bright spike they had recorded showed only one tenth the luminosity to be expected from stars with that much mass, suggestive of a nonluminous object.

Both groups, in their separate reports, pointed out that the existence of a black hole had not been proved, and that other explanations were possible. Nevertheless, as the Palomar astronomers put it:

M 87 is at present probably the most plausible case
for a massive black hole in a galaxy nucleus. The

existing data are pressing upon the current limits for ground-based observations, and are not likely to be improved upon in the near future. Probably the best hope for a dramatic improvement in the data lies with the Space Telescope.

The latter is to be launched in 1983 and, the astronomers pointed out, it should increase tenfold the amount of detail that can be seen in M 87 and other such galaxies.

To inhabitants of the planet Earth, orbiting a star in the Milky Way Galaxy, the most exciting news concerns recent observations indicating that a fierce, but comparatively well behaved "machine" is at work in the nucleus of our own galaxy. For example, Kenneth I. Kellermann and his colleagues at the National Radio Astronomy Observatory have found evidence for an extremely compact object there. They conducted interferometry (VLBI) measurements using the 37-meter Haystack dish in Massachusetts; their own observatory's 43-meter antenna in West Virginia; and NASA's 64-meter antenna at Goldstone, California, designed for tracking distant spacecraft. From the recorded emissions they calculated that the radio source known as Sagittarius A West (presumably in the core) is no more than 200 astronomical units wide (the diameter of the orbit of Pluto, the outermost planet, is 80 such units). A quarter of the radiation comes from a central region only 10 astonomical units across. Since earlier observations at larger wavelengths indicated larger widths, it was suspected that scattering of the waves en route (chiefly at longer wavelengths) made the source look larger than it really is and that, if scanned at even shorter wavelengths, it would prove still more compact. The same spot also appears to be a source of intense infrared emissions.

The radio emissions are not in a class with those from the powerful radio galaxies, and it is not certain that their manner of generation is similar. After suggesting such explanations for the source as a tight bunching of stars or a supernova, the authors cited suggestions of a black hole at the center of the galaxy to explain the various phenomena being observed there. They noted the proposal of Lynden-Bell with regard to the energy machine in the cores of galaxies and his suggestion that detection of a very compact radio source would be a test of the hypothesis.

On April 26, 1976, the Navy group, in a rocket observation of the core region, were able to identify four discrete sources of intense X-ray emission. None, however, coincided with the radio source Sagittarius A West or other radio or infrared sources there, implying that the X rays may not be coming from the central "machine." Nevertheless, their results, they reported, "are compatible with the nucleus being a massive black hole, around which accretion should produce temperatures too low for X-ray emission." This, they noted, fit a concept of Lynden-Bell and Rees in which the ultraviolet radiation from the nucleus heats a surrounding envelope of dust. This hot dust then generates the observed infrared radiation.

Even though the core of our galaxy seems well behaved at the moment, there is evidence of periodic explosions. As pointed out by Friedman, "A doughnut-shaped molecular cloud, containing the mass of a hundred million suns, surrounds the nucleus and is blowing outward like a gigantic smoke ring. Its diameter is about 1,600 light years and it is expanding at a speed of about 260,000 miles per hour, implying a great explosion only a million years ago. Still farther out is a rapidly rotating disk of hydrogen gas, expanding at a rate of a million miles per hour, signifying an even earlier explosion."

Jan H. Oort of the Netherlands, dean of European radio astronomers, summarizing what has been observed in the core, said in 1977 that an "ultracompact" radio source far smaller than the solar system—a grain of sand on the scale of the Milky Way system—"is presumably the actual center of the Galaxy. It may have a mass in the order of five million solar masses and there is a suspicion that it may contain the 'engine' responsible for the many explosive phenomena observed throughout the central region."

Does this mean, then, that The Machine blowing apart radio galaxies and causing quasars to shine with incredible brilliance lies smoldering in the core of our own star system? That now seems a strong possibility.

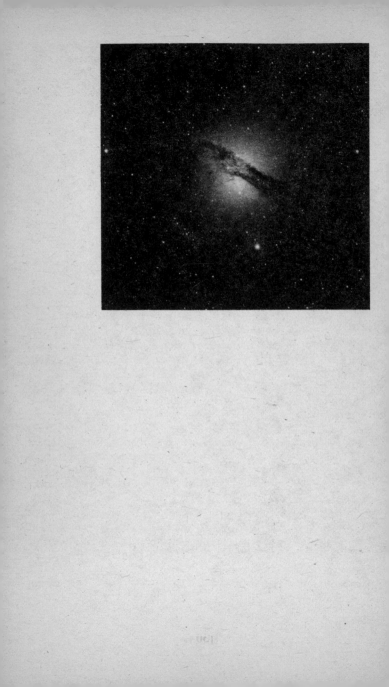

17
Epilogue

Star after star from heaven's high arch shall rush,
Suns sink on suns, and systems systems crush,
Headlong, extinct, in one dark centre fall,
And death, and night, and chaos mingle all!
ERASMUS DARWIN*

The story told in this book cannot end here. We do not know
for sure whether the universe is open or closed (for example,
by countless black holes). We do not know whether black
holes, as such, exist, and whether within them singularities
provide links to other realms of space and time. Perhaps
"cosmic censorship" will forever deny us answers to these
questions.

If the universe expands forever, the stars, one by one,
will collapse into white dwarfs, neutron stars, and black holes
(or some other form of superdense state). The white dwarfs
will cool into black dwarfs. The pulsars will radiate away
their energy and run down. The end will be universal darkness.

If, as many would prefer on philosophical grounds, the
universe stops expanding and collapses, perhaps to rebound
into a new universe as part of a cycle without beginning or
end, the prospects for anyone alive at the time of collapse are
no more appealing.

Even though we do not know which of these fates awaits

*(Grandfather of Charles), *Economy of Vegetation*, Canto IV (1792).

the universe or whether black holes exist precisely in the manner predicted by relativity, the observations of the past few years leave little doubt that massive gravitational collapse does occur and that—presumably in this manner—energy is being generated on a scale far beyond anything previously considered possible. If supermassive collapse does not end in a black hole, it must end in something else. Very likely in Cyg X-1, M 87, and elsewhere we are observing such objects. Thus gravity, the weakest force we know and the only one with which we have intimate daily experience, has proved the most powerful of all energy sources. In large-scale collapse it can reissue, on a catastrophic scale, some of the energy that blew the universe apart in the first place.

We know as well that conditions that should produce black holes, according to our limited knowledge of physics, do exist: stars of such enormous mass that, when their internal fires go out, no known form of resistance can halt their collapse, at lightning speed, past white-dwarf density and past that of a neutron star to infinite density in zero volume. From experiment and observation we know that, at least under less extreme circumstances, the predictions of relativity concerning the effects of gravity in bending space and slowing time are valid. Unless these effects are modified in the death of a supermassive star, collapse from the viewpoint of someone falling into the hole proceeds rapidly to a singularity, where space and time vanish (or go "somewhere else"). To an external observer the effect of the hole's intense gravity is to slow time essentially to a halt, so that the collapse (if it could be seen) would seem to linger.

Since no light or other emission could escape such an object (except in the special manner predicted by Hawking), detection of black holes is extremely difficult—as demonstrated in the debate over Cyg X-1. Yet, as recounted in this book, there is no question that less massive collapses occur, producing white dwarfs and neutron stars (evident as radio pulsars or in the two-star systems where they pulse in X rays). These objects manifest conditions of density, spin rate, magnetism, and (in the X-ray pulsars) temperature so extreme that the infinities of a black hole become more plausible. As a consequence astrophysicists of the highest repute are now persuaded that black holes exist, just as others, equally qualified, are skeptical.

If we learn what really happens in the collapse of large assemblages of matter, such as the giant stars, this may answer the more fundamental question of what would occur if the entire universe collapses. Would it "bounce" into a new universe or vanish in a singularity? Perhaps we will uncover a new level of physical law. Edward Harrison, the radio astronomer, has cited the revolution that occurred in physics when it was found that the rules applicable to objects like planets, rocks, and grains of sand break down on the atomic level and that new rules—quantum mechanics—had to be invented. Will the same be true, he has asked, of the opposite extreme—the very large, the very massive? Or, in finally understanding the smallest things, such as quarks, will we guess what happens in collapse of such supermassive objects as giant stars or the universe itself?

It is a special joy of scientific inquiry that no truth is absolute. No theory that attempts to describe nature can be complete. In the past few years new observing devices—giant optical and radio telescopes, instruments lifted by rockets or placed in earth orbit—have revealed a host of new wonders including pulsars, quasars, X-ray binaries, and even the "glow" from the primordial fireball (or from the starlight of universes past or future). Not even the wildest imaginations of generations gone before have conjured up such concepts— not even the ancient Greeks, with their vision of Zeus hurling thunderbolts, nor the Icelanders, with their legend of Ymir, the giant whose skull became the sky, his bones the rocks of the earth, and his blood the sea.

Although man has gazed in awe upon the heavens for hundreds of thousands of years, only in the lifetime of the present generation has it been possible to look far enough to reach close to "the beginning." Each door that we open on the universe, however, seems to reveal not final answers but more doors to seek to open. With our satellite-borne instruments and multitudinous other new ways of looking at the universe we are bound to find new doors. It may well be that the search will never end so long as our descendants look skyward.

Selected References

In many ways this book can be considered an ongoing scientific mystery story. The following references are for those who would like to see some of the original clues. Included are sources from which most of the quotations were taken as well as sources of special historical interest, apart from those that are generally inaccessible. Where appropriate, attention is drawn to more extensive bibliographies. Unless otherwise noted, references to Soviet publications are to the translated editions (in English).

Chapter 1 · June 30, 1908

Astapovitch, I. S. "Air waves caused by the fall of the meteorite on 30th June, 1908, in Central Siberia," *Quarterly Journal of the Royal Meteorological Society,* Vol. 60 (1934), pp. 493–504.

Baxter, John, and Atkins, Thomas. *The Fire Came By,* Garden City, N.Y.: Doubleday & Company, 1976; pp. 161–65 list many citations on which this chapter was based.

Beasley, William H., and Tinsley, Brian A. "Tungus event was not caused by a black hole," *Nature,* Vol. 250 (1974), pp. 555–56.

Ben-Menahem, Ari. "Source parameters of the Siberian explosion of June 30, 1908, from analysis and synthesis of seismic signals at four stations," *Physics of the Earth and Planetary Interiors,* Vol. 11 (1975), pp. 1–34.

Brown, John C., and Hughes, David W. "Tunguska's comet and non-thermal ^{14}C production in the atmosphere," *Nature,* Vol. 268 (1977), pp. 512–14.

Cowan, Clyde; Atluri, C. R.; and Libby, W. F. "Possible anti-matter content of the Tunguska meteor of 1908," *Nature,* Vol. 206 (1965), pp. 861–65.

Fesenkov, V. G. "On the origin of comets and their importance for the cosmogony of the solar system," in *The Motion, Evolution of Orbits, and Origin of Comets,* Chebotarev, G. A., et al., editors, IAU Symposium Nr. 45, D. Reidel, Dordrecht (1972), pp. 409–12.

Gentry, Robert V. "Anti-matter content of the Tunguska meteor," *Nature,* Vol. 211 (1966), pp. 1071–72.

Hawking, Stephen. "Gravitationally collapsed objects of very low mass," *Monthly Notices of the Royal Astronomical Society,* Vol. 152 (1971), pp. 75–78.

Hunt, J. N.; Palmer, R.; and Penney, Sir William. "Atmospheric waves caused by large explosions," *Philosophical Transactions of the Royal Society of London,* Series A, Vol. 252 (1960), pp. 275–315.

Jackson, A. A. IV, and Ryan, Michael P., jun. "Was the Tungus event due to a black hole?" *Nature,* Vol. 245 (1973), pp. 88–89.

Jones, G. H. S. "High-explosive analogue of the Tunguska event," *Nature,* Vol. 267 (1977), p. 605. See also commentary by B. W. Augenstein, *Nature,* Vol. 269 (1977), p. 355.

Kresák, L. "The Tunguska object: a fragment of the comet Encke?" Bulletin, Astronomical Institutes of Czechoslovakia, Vol. 29 (1978), pp. 129–34.

Krinov, E. L. "Commentary on Kulik's The Tunguska Meteorite," *Source Book in Astronomy 1900–1950,* ed. Harlow Shapley. Cambridge, Mass.: Harvard University Press (1960), pp. 79–81.

———. *Giant Meteorites.* Pergamon (New York, 1966).

Kulik, L. A. "On the problem of the fall of the Tunguska Meteorite 1908" (in Russian), *Doklady Akademmy Nauk SSSR,* Vol. 23 (1927), pp. 399–402.

Rich, Vera. "The 70-year-old mystery of Siberia's big bang," *Nature,* Vol. 274 (1978), p. 207.

———. "Tungus expert discovers biofields," *ibid.,* p. 305.

Seuss, Hans E. "Secular variations of the cosmic-ray-produced carbon 14 in the atmosphere and their implications," *Journal of Geophysical Research,* Vol. 70 (1965), pp. 5937–52.

Whipple, F. J. W. "The great Siberian meteor and the waves it produced," *Quarterly Journal of the Royal Meteorological Society,* Vol. 56 (1930), pp. 287–304.

Whipple, Fred L. "Comments on the 1908 Tunguska explosion" in "Do comets play a role in galactic chemistry and gamma-ray bursts?" *Astronomical Journal,* Vol. 80, (1975), p. 530.

Zolotov, A. V. "The possibility of 'thermal' explosion and the structure of the Tungus meteorite," *Soviet Physics—Doklady*, Vol. 12 (1967), pp. 101–4.

Zotkin, I. T., and Tsikulin, M. A. "Simulation of the explosion of the Tungus meteorite," *Soviet Physics—Doklady*, Vol. 11 (1966), pp. 183–86.

Chapter 2 · Ten Tons per Cubic Inch

Chandrasekhar, S. "The density of white-dwarf stars," *The London Edinburgh and Dublin Philosophical Magazine and Journal of Science*, Vol. 11 (1931), pp. 592–96.

————. "The highly collapsed configurations of a stellar mass," *Monthly Notices of the Royal Astronomical Society*, Vol. 91 (1931), pp. 456–66.

————. "The maximum mass of ideal white dwarfs," *The Astrophysical Journal*, Vol. 74 (1931), pp. 81–82.

————. "The increasing role of general relativity in astronomy," Halley Lecture for 1972, *Observatory*, Vol. 92 (1972), pp. 160–74. Contains citations from A. S. Eddington and E. A. Milne.

Compton, A. H. *Atomic Quest*. New York: Oxford University Press, 1956, p. 144.

Eddington, A. S. *The Internal Constitution of the Stars*. New York: Cambridge University Press, 1926.

Newton, Isaac. *Optiks*, quoted by J. H. Jeans in *Source Book in Astronomy 1900–1950*, ed. Harlow Shapley. Cambridge, Mass.: Harvard University Press, 1960, p. 343.

Thackray, Arnold. " 'Matter in a nut-shell': Newton's *Optiks* and Eighteenth-century chemistry," *AMBIX—The Journal of the Society for the Study of Alchemy and Early Chemistry*, Vol. 15 (1968), pp. 29–53.

Chapter 3 · Brighter Than a Hundred Billion Stars

Baade, W., and Zwicky, F. "Supernovae and cosmic rays," abstract from December 1933 meeting of American Physical Society, in *Physical Review*, Vol. 45 (1934), p. 138.

————. "On Super-novae," *Proceedings of the National Academy of Sciences*, Vol. 20 (1934), pp. 254–59, and next article, by same authors, "Cosmic rays from super-novae," pp. 259–63.

Gamow, George. *Structure of Atomic Nuclei and Nuclear Transformations*. New York: Oxford University Press, 1937, pp. 234–38.

Landau, Lev. "On the theory of stars" (in English), *Physikalische Zeitschrift Der Sowjetunion*, Vol. 1 (1932), pp. 285–88.

———. "Origin of stellar energy," *Nature*, Vol. 141 (1938), pp. 333–34.

Zwicky, F. "On collapsed neutron stars," *Astrophysical Journal*, Vol. 88 (1938), pp. 522–25.

———. "On the theory and observation of highly collapsed stars," *Physical Review*, Vol. 55 (1939), pp. 726–43.

Chapter 4 · The Warp of Space and Time

Chandrasekhar, S. "Verifying the theory of relativity," *Notes and Records of the Royal Society of London*, Vol. 30 (Jan. 1976), pp. 249–60. Contains citations from Dyson, Eddington, Jeans, and Silberstein.

Clark, Ronald W. *Einstein—The Life and Times*. New York: Thomas Y. Crowell Company, 1971, pp. 228–29.

Darwin, C. G. "The clock paradox in relativity," *Nature*, Vol. 180 (1957), pp. 976–77.

Einstein, A. "Die Relativitäts Theorie," in "Sitzungsberichte von 1911," *Vierteljahresschrift, Naturforschende Gesellschaft, Zürich*, Vol. 56 (1911), pp. 1–14. See also in this volume "Diskusion—Sitzung vom 16. Januar 1911."

Hafele, J. C., and Keating, Richard E. "Around-the-world atomic clocks: predicted relativistic time gains" and "Around-the-world atomic clocks: observed relativistic time gains," *Science*, Vol. 177 (1972), pp. 166–68 and 168–70.

Hoffmann, Banesh, with Dukas, Helen. *Albert Einstein—Creator and Rebel*. New York: Viking Press, 1972.

Holton, Gerald J. "Time dilation and clock problems" in "Resource letter SRT 1 on special relativity theory," *American Journal of Physics*, Vol. 30 (1962), pp. 467–68.

Kaufmann, William J. *The Cosmic Frontiers of General Relativity*. Boston: Little, Brown, 1977.

Sachs, Mendel. "A resolution of the clock paradox," *Physics Today*, Vol. 24 (Sept. 1971), pp. 23–29. A reply by James Terrell appears in the Jan. 1972 issue, p. 9.

Chapter 5 · The Black-Hole Idea Born

Eddington, A. S. *The Internal Constitution of the Stars*. London: Cambridge University Press, 1926, p. 6.

————. Obituary notice on K. Schwarzschild, cited by S. Chandrasekhar, "Verifying the theory of relativity," op. cit., pp. 257–58.

Einstein, A. Letter to K. Schwarzschild, partially quoted by S. Chandrasekhar, ibid., p. 259.

Laplace, P. S. *Exposition du Système du Monde*, 2nd ed., Paris: "An VII," pp. 546–49.

Oppenheimer, J. R., and Serber, Robert. "On the stability of stellar neutron cores," *Physical Review*, Vol. 54 (1938), p. 540.

Oppenheimer, J. R., and Volkoff, G. M. "On massive neutron cores," *Physical Review*, Vol. 55 (1939), pp. 374–81.

Oppenheimer, J. R., and Snyder, H. "On continued gravitational contraction," *Physical Review*, Vol. 56 (1939), pp. 455–59.

Schwarzschild, K. "Über das Gravitationsfeld eines Massenpunktes nach der Einsteinschen Theorie," *Sitzungsberichte der Königlich Preussischen Akademie der Wissenschaften* (1916), pp. 189–96.

Chapter 6 · Cosmic Distances and Cosmic Explosions

Hoyle, F., and Fowler, W. A. "On the nature of strong radio sources," *Monthly Notices of the Royal Astronomical Society*, Vol. 125 (1963), pp. 169–76.

————. "Nature of strong radio sources," *Nature*, Vol. 197 (1963), pp. 533–35.

Hoyle, F.; Fowler, William A.; Burbidge, G. R.; and Burbidge, E. Margaret. "On relativistic astrophysics," *Astrophysical Journal* Vol. 139 (1964), pp. 909–28.

Pfeiffer, J. *The Changing Universe*, New York: Random House, 1956, pp. 107–11.

Robinson, Ivor; Schild, Alfred; and Schucking, E. L. (eds.) *Quasi-stellar Sources and Gravitational Collapse—Including the Proceedings of the First Texas Symposium on Relativistic Astrophysics*, Chicago, Ill.: University of Chicago Press, 1965, pp. xi–xvii.

Spitzer, Lyman, Jr., and Baade, Walter. "Stellar populations and collisions of galaxies," *Astrophysical Journal*, Vol. 113 (1951), pp. 413–18.

Alfvén, Hannes, and Elvius, Aina. "Antimatter, quasi-stellar objects, and the evolution of galaxies," *Science,* Vol. 164 (1969), pp. 911–17.

Alpher, Ralph A., and Herman, Robert. "Evolution of the universe," *Nature,* Vol. 162 (1948), pp. 774–75.

Anderson, B.; Donaldson, W.; Palmer, H. P.; and Rowson, B. "Observations of the quasi-stellar and other sources with an interferometer of resolving power 0.4 sec. of arc," *Nature,* Vol. 205 (1965), pp. 375–76.

Apparao, K. M. V.; Bignami, G. F.; Maraschi, L.; Helmken, H.; Margon, B.; Hjellming, R.; Bradt, H. V.; and Dower, R. G. "4U0241+61: A luminous low-redshift QSO," *Nature,* vol. 273 (1978), pp. 450–53.

Burbidge, G. R., and Burbidge, E. M. "Quasi-stellar objects—a progress report," *Nature,* Vol. 224 (1969), pp. 21–24.

Burbidge, G. R., and Hoyle, F. "On the space distribution of identified quasi-stellar objects and radio galaxies," *Nature,* Vol. 216 (1967), pp. 351–52.

Chandrasekhar, S., footnote to article by M. Schmidt, *Astrophysical Journal,* Vol. 162 (1970), p. 371.

Dicke, R. H.; Peebles, P. J. E.; Roll, P. G.; and Wilkinson, D. T. "Cosmic black-body radiation," *Astrophysical Journal,* Vol. 142 (1965), pp. 414–19.

Disney, Michael J., and Véron, Philippe. "BL Lacertae Objects," *Scientific American,* Vol. 237 (Aug. 1977), pp. 32–39.

Eachus, Lola J., and Liller, William. "Photometric histories of QSOs: 3C 279, the most variable and possibly most luminous QSO yet studied," *Astrophysical Journal,* Vol. 200 (1975) pp. L61–L62.

Field, G. B. "Quasi-stellar radio sources as spherical galaxies in the process of formation," *Nature,* Vol. 202 (1964), pp. 786–87.

Gottlieb, Elaine W., and Liller, William. "The historical light curve of PKS 2134+004, a highly luminous QSO," *Astrophysical Journal,* Vol. 222 (1978), pp. L1–L2.

Greenstein, Jesse L. "Quasi-stellar radio sources," *Scientific American,* Vol. 209 (Dec. 1963), pp. 54–62.

Greenstein, Jesse L., and Matthews, Thomas A. "Red-shift of the unusual radio source 3C 48," *Nature,* Vol. 197 (1963), pp. 1041–42.

Hazard, C.; Mackey, M. B.; and Nicolson, W. "Additional identifications of radio sources with star-like objects," *Nature*, Vol. 202 (1964), pp. 227–28.

Hazard, C.; Mackey, M. B.; and Shimmins, A. J. "Investigation of the radio source 3C 273 by the method of lunar occultations," *Nature*, Vol. 197 (1963), pp. 1037–39.

Hoyle, F., and Burbidge, G. R. "Relation between the red-shifts of quasi-stellar objects and their radio and optical magnitudes," *Nature*, Vol. 210 (1966), pp. 1346–47.

Hoyle F.; Burbidge, G. R.; and Sargent, W. L. W. "On the nature of the quasi-stellar sources," *Nature*, Vol. 209 (1966), pp. 751–53.

McCrea, W. H. "Quasars: Rapid light fluctuations," *Science*, Vol. 157 (1967), pp. 400–2.

Metz, William D. "New light on quasars: Unraveling the mystery of BL Lacertae," *Science*, Vol. 200 (1978), pp. 1031–33.

Morrison, Philip. "Resolving the mystery of the quasars?" *Physics Today*, Vol. 26 (Mar. 1973), pp. 23–29.

Oke, J. B. "Absolute energy distribution in the optical spectrum of 3C 273," *Nature*, Vol. 197 (1963), pp. 1040–41.

Penzias, A. A., and Wilson, R. W. "A measurement of excess antenna temperature at 4080 Mc/s," *Astrophysical Journal*, Vol. 142 (1965), pp. 419–21.

Robinson, Ivor, et al. *Quasi-stellar Sources and Gravitational Collapse—Including Proceedings of the First Texas Symposium on Relativistic Astrophysics*, op. cit. Contains a number of the papers presented as well as the invitation circulated beforehand.

Sandage, Allan. "The existence of a major new constituent of the universe: The quasi-stellar galaxies," *Astrophysical Journal*, Vol. 141 (1965), pp. 1560–78.

Schmidt, M. "3C 273: A star-like object with large red-shift," *Nature*, Vol. 197 (1963), p. 1040.

———. "Quasi-stellar objects," *Science Journal*, Vol. 2 (Oct. 1966), pp. 77–83.

Terrell, James. "Quasi-stellar diameters and intensity fluctuations," *Science*, Vol. 145 (1964), pp. 918–19.

Weinberg, Steven. *The First Three Minutes: A Modern View of the Origin of the Universe*. New York: Basic Books, 1977. Includes a detailed account of the prediction and discovery of the residual fireball radiation.

Wheeler, John Archibald. "Geons," *Physical Review*, Vol. 97 (1955), pp. 511–36.

Chapter 8 · The "Little Green Men"

Baade, Walter. "The Crab Nebula," *Astrophysical Journal*, Vol. 96 (1942), pp. 188–98.

Bowyer, S.; Byram, E. T.; Chubb, T. A.; and Friedman, H. "X-ray sources in the galaxy," *Nature*, Vol. 201 (1964), pp. 1307–8 (see also following article by Donald C. Morton on the neutron-star hypothesis of Hong-yee Chiu).

Burnell, S. Jocelyn Bell. "Petit Four," *Eighth Texas Symposium on Relativistic Astrophysics, Annals of the New York Academy of Sciences*, Vol. 302 (1977), pp. 685–89.

Cocke, W. J.; Disney, M. J.; and Taylor, D. J. "Discovery of optical signals from pulsar NP 0532," *Nature*, Vol. 221 (1969), pp. 525–27.

Duthie, J. G.; Sturch, C.; and Hafner, E. M. "Optical pulse of a periodic radio star," *Science*, Vol. 160 (1968), pp. 415–16.

Fritz, G.; Henry, R. C.; Meekins, J. F.; Chubb, T. A.; and Friedman, H. "X-ray pulsar in the Crab Nebula," *Science*, Vol. 164 (1969), pp. 709–12.

Gold, T. "Rotating neutron stars as the origin of the pulsating radio sources," *Nature*, Vol. 218 (1968), pp. 731–32.

———. "Rotating neutron stars and the nature of pulsars," *Nature*, Vol. 221 (1969), pp. 25–27.

———. "Thomas Gold talks about pulsars—the key to cosmic rays?" *Scientific Research* (Cornell University) (June 9, 1969), unsigned, pp. 32–36.

Green, Louis C. "Pulsars today" (Parts I and II), *Sky and Telescope*, Vol. 40 (Nov. and Dec. 1970), pp. 260–62, 357–60.

———. "Starquakes: Have they been observed?" *Sky and Telescope*, Vol. 41 (Feb. 1971), pp. 76–79.

Gunn, James E., and Ostriker, Jeremiah P. "Magnetic dipole radiation from pulsars," *Nature*, Vol. 221 (1969), pp. 454–56.

Hewish, Antony. "Pulsars," *Scientific American*, Vol. 219 (Oct. 1968), pp. 25–35.

———. "Pulsars and high-density physics" (Nobel Lecture), *Science*, Vol. 188 (1976), pp. 1079–83.

Hewish, A.; Bell, S. J.; Pilkington, J. D. H.; Scott, P. F.; and Collins, R. A. "Observation of a rapidly pulsating radio source," *Nature*, Vol. 217 (1968), pp. 709–13.

Nather, R. E.; Warner, B.; and Macfarlane, M. "Optical pulsations in the Crab Nebula pulsar," *Nature*, Vol. 221 (1969), pp. 527–29.

Staelin, David H., and Reifenstein, Edward C. III. "Pulsating radio sources near the Crab Nebula," *Science*, Vol. 162 (1968), pp. 1481–83.

Wade, Nicholas, "Discovery of pulsars: A graduate student's story," *Science*, Vol. 189 (1975), pp. 358–64.

Warnow, Joan M. (ed.) "Moments of discovery: Optical pulsars," *The Living History of Physics and the Human Dimension of Science*, Center for History of Physics, American Institute of Physics.

Chapter 9 · *Uhuru* and the X-ray Sky

Bowyer, S.; Byram, E. T.; Chubb, T. A.; and Friedman, H. "X-ray sources in the galaxy," *Nature*, Vol. 201 (1964), pp. 1307–8.

———. "Lunar occultation of X-ray emission from the Crab Nebula," *Science*, Vol. 146 (1964), pp. 912–17.

———. "Cosmic X-ray sources," *Science*, Vol. 147 (1965), pp. 394–98.

Byram, E. T.; Chubb, T. A.; and Friedman, H. "Cosmic X-ray sources, galactic and extragalactic," *Science*, Vol. 152 (1966), pp. 66–71.

Clark, G.; Garmire, G.; Oda, M.; Wada, M.; Giaconni, R.; Gursky, H.; and Waters, J. R. "Positions of three cosmic X-ray sources in Scorpio and Sagittarius," *Nature*, Vol. 207 (1965), pp. 584–87.

Friedman, Herbert, "Rocket observations of the ionosphere," *Proceedings of the Institute of Radio Engineers*, Vol. 47 (Feb. 1959), pp. 272–80.

———. "X-ray astronomy," *New Scientist*, Vol. 28 (Dec. 30, 1965), pp. 904–6.

———. "Rocket astronomy," *Annals of the New York Academy of Sciences*, Vol. 198 (1972), pp. 267–73.

———. "Reminiscences of thirty years of space research," *NRL Report 8113*, Naval Research Laboratory, Washington, D.C. (Aug. 1977).

———. *The Amazing Universe*, Washington, D.C.: National Geographic Society, 1975, p. 117.

Fritz, G.; Henry, R. C.; Meekins, J. F.; Chubb, T. A.; and Friedman, H. "X-ray pulsar in the Crab Nebula," *Science*, Vol. 164 (1969), pp. 709–12.

Giacconi, Riccardo; Gursky, Herbert; and Paolini, Frank R. "Evidence for X-rays from sources outside the solar system," *Physical Review Letters*, Vol. 9 (1962), pp. 439–43.

Giacconi, R.; Gursky, H.; and Waters, J. R. "Spectral data from the cosmic X-ray sources in Scorpius and near the galactic centre," *Nature*, Vol. 207 (1965), pp. 572–75.

Giacconi, R.; Kellogg, E.; Gorenstein, P.; Gursky, H.; and Tananbaum, H. "An X-ray scan of the galactic plane from *Uhuru*," *Astrophysical Journal*, Vol. 165 (1971), pp. L27–L35.

Giacconi, R.; Murray, S.; Gursky, H.; Kellogg, E.; Schreier, E.; and Tananbaum, H. "The *Uhuru* catalogue of X-ray sources," *Astrophysical Journal*, Vol. 178 (1972), pp. 281–308.

Giacconi, Riccardo, and Gursky, Herbert (eds.). *X-ray Astronomy*, Dordrecht-Holland: D. Reidel, 1974. See especially Introduction by Giacconi, pp. 1–23, with extensive references.

Gursky, Herbert; Giacconi, Riccardo; Paolini, Frank R.; and Rossi, Bruno B. "Further evidence for the existence of galactic X rays, *Physical Review Letters*, Vol. 11 (1963), pp. 530–35.

Gursky, H.; Giacconi, R.; Gorenstein, P.; Waters, J. R.; Oda, M.; Bradt, H.; Garmire, G.; and Sreekantan, B. V. "A measurement of the location of the X-ray source Sco X-1," *Astrophysical Journal*, Vol. 146 (1966), pp. 310–16.

Johnson, Hugh M., and Stephenson, C. B. "A possible old nova near Sco X-1," *Astrophysical Journal*, Vol. 146 (1966), pp. 602–4.

Oda, Minoru. "High-resolution X-ray collimator with broad field of view for astronomical use," *Applied Optics*, Vol. 4 (1965), p. 143.

Oda, M.; Clark, G.; Garmire, G.; Wada, M.; Giacconi, R.; Gursky, H.; and Waters, J. "Angular sizes of the X-ray sources in Scorpio and Sagittarius," *Nature*, Vol. 205 (1965), pp. 554–55.

Sandage, A. R.; Osmer, P.; Giacconi, R.; Gorenstein, P.; Gursky, H.; Waters, J.; Bradt, H.; Garmire, G.; Sreekantan, B. V.; Oda, M.; Osawa, K.; and Jugaku, J. "On the optical identification of Sco X-1," *Astrophysical Journal*, Vol. 146 (1966), pp. 316–21.

Tousey, Richard. "Highlights of twenty years of optical space research," *Applied Optics*, Vol. 6 (1967), pp. 2044–77.

Chapter 10 · The "Demon Bird of Satan"

Ananthakrishnan, S., and Ramanathan, K. R. "Effects on the lower inosphere of X-rays from Scorpius XR-1," *Nature*, Vol. 223 (1969), pp. 488–89.

Bahcall, John N., and Bahcall, Neta A. "The period and light curve of HZ Herculis," *Astrophysical Journal,* Vol. 178 (1972), pp. L1–L4.

Bolton, C. T. "Identification of Cygnus X-1 with HDE 226868," *Nature,* Vol. 235 (1972), pp. 271–73.

————. "Dimensions of the binary system HDE 226868—Cygnus X-1," *Nature Physical Science,* Vol. 240 (1972), pp. 124–26.

Boynton, P. E. "The Clockwork Wonder," in *Physics and Astrophysics of Neutron Stars and Black Holes,* eds. R. Giacconi and R. Rufino. Amsterdam: North Holland Publishing Co., 1978. Proceedings of the International School of Physics "Enrico Fermi" Course LXV, Società Italiana di Fisica, 14–28, July 1975.

Bradt, H., and Giacconi, R. (eds.). *X- and Gamma-ray Astronomy,* Dordrecht-Holland: D. Reidel, 1973, IAU Symposium held in Madrid May 11–13, 1972. Contains observational and theoretical reports on X-ray observations in space.

Braes, L. L. E., and Miley, G. K. "Detection of radio emission from Cygnus X-1," *Nature,* Vol. 232 (1971), p. 246.

————. "Further radio observations of Cygnus X-1," *Nature Physical Science,* Vol. 235 (1972), p. 147.

————. "Radio detection of Cygnus X-3," *Nature,* Vol. 237 (1972), p. 506.

————. "Another correlated X-ray-radio transition in Cygnus X-1," *Nature,* Vol. 264 (1976), pp. 731–32.

Byram, E. T.; Chubb, T. A.; and Friedman, H. "Cosmic X-ray sources, galactic and extragalactic," *Science,* Vol. 152 (1966), pp. 66–71.

Chodil, G.; Mark, Hans; Rodrigues, R.; Steward, F.; Swift, C. D.; Hiltner, W. A.; Wallerstein, George; and Manner, Edward J. "Spectral and location measurements of several cosmic X-ray sources including a variable source in Centaurus," *Physical Review Letters,* Vol. 19 (1967), pp. 681–83.

Coe, M. J.; Engel, A. R.; and Quenby, J. J. "Anti-correlated hard and soft X-ray intensity variations of the black-hole candidates Cyg X-1 and A0620-00," *Nature,* Vol. 259 (1976), pp. 544–45.

Cominsky, L.; Clark, G. W.; Li, F.; Mayer, W.; and Rappaport, S. "Discovery of 3.6-s X-ray pulsations from 4U0115 + 63," *Nature,* Vol. 273 (1978), pp. 367–69.

Cooke, B. A., and Pounds, K. A. "Further high-sensitivity X-ray sky survey from the Southern Hemisphere," *Nature Physical Science,* Vol. 229 (1971), pp. 144–47.

Elsner, R. F., and Lamb, F. K. "Accretion flows in the magnetospheres of Vela X-1, A0535+26 and Her X-1," *Nature*, Vol. 262 (1976), pp. 356–60.

Forman, William; Jones, Christine A.; and Liller, William. "Optical studies of *Uhuru* sources. III. Optical variations of the X-ray eclipsing system HZ Herculis," *Astrophysical Journal*, Vol. 177 (1972), pp. L103–L107.

Friedman, Herbert. "Cosmic X-ray sources: A progress report," *Science*, Vol. 181 (1973), pp. 395–407.

Giacconi, R.; Gursky, H.; Kellogg, E.; Schreier, E.; and Tananbaum, H. "Discovery of periodic X-ray pulsations in Centaurus X-3 from *Uhuru*," *Astrophysical Journal*, Vol. 167 (1971), pp. L67–L73.

Giacconi, Riccardo. "Binary X-ray sources," in *Gravitational Radiation and Gravitational Collapse*, ed. C. DeWitt-Morette. Dordrecht/Boston, IAU, 1974, pp. 147–80.

Gottlieb, Elaine W.; Wright, Edward L.; and Liller, William. "Optical studies of *Uhuru* sources. XI. A probable period for Scorpius X-1 = V818 Scorpii," *Astrophysical Journal*, Vol. 195 (1975), pp. L33–L35.

Gregory, P. C. "Large outburst of Cygnus X-3," *Nature*, Vol. 239 (1972), followed by two detailed reports, pp. 439–46.

Heise, J.; Brinkman, A. C.; Schrijver, J.; Mewe, R.; den Boggende, A.; and Gronenschild, E. "X-ray observations of Cyg X-1 with ANS," *Nature*, Vol. 256 (1975), pp. 107–8.

Hjellming, R. M., and Wade, C. M. "Radio emission from X-ray sources," *Astrophysical Journal*, Vol. 168 (1971), pp. L21–L24.

Hjellming R. M.; Hermann, M.; and Webster, E. "Radio observation of Cygnus X-3," *Nature*, Vol. 237 (1972), pp. 507–8.

Hjellming, R. M.; Gibson, D. M.; and Owen, F. N. "Another major change in the radio source associated with Cyg X-1," *Nature*, Vol. 256 (1975), pp. 111–12.

Holt, S. S.; Boldt, E. A.; Kaluzienski, L. J.; and Serlemitsos, P. J. "Observation of a new transition in the emission from Cyg X-1," *Nature*, Vol. 256 (1975), pp. 108–9.

Holt, S. S.; Boldt, E. A.; Serlemitsos, P. J.; Kaluzienski, L. J.; Pravdo, S. H.; Peacock, A.; Elvis, M.; Watson, M. G.; and Pounds, K. A. "Evidence for a 17-d periodicity from Cyg X-3," *Nature*, Vol. 260 (1976), pp. 592–94.

Jones, C.; Giacconi, R.; Forman, W.; and Tananbaum, H. "Observations of Circinus X-1 from *Uhuru*," *Astrophysical Journal*, Vol. 191 (1974), pp. L71–L74.

Kristian, J.; Brucato, R.; Visvanathan, N.; Lanning, H.; and Sandage, A. "On the optical identification of Cygnus X-1, *Astrophysical Journal*, Vol. 168 (1971), pp. L91–92.

Lamb, D. Q.; and Sorvari, J. M. "HZ Her as a possible optical pulsar," *Circular 2422* (July 17, 1972), IAU Central Bureau for Astronomical Telegrams.

Liller, William. "The story of AM Herculis," *Sky and Telescope*, Vol. 53 (May 1977), pp. 351–54.

Lyutyin, V. M.; Sunyaev, R. A.; and Cherepashchuk, A. M. "Nature of the optical variability of HZ Herculis (Her X-1) and BD + 34° 3815 (Cyg X-1), *Soviet Astronomy*, Vol. 17 (1973), pp. 1–6.

Mason, Keith O.; Hawkins, Frederick J.; and Sanford, Peter W. "X-ray absorption events in Cygnus X-1 observed with *Copernicus*," *Astrophysical Journal*, Vol. 192 (1974), pp. L65–L69.

McClintock, J. E.; Rappaport, S.; Joss, P. C.; Bradt, H.; Buff, J.; Clark, G. W.; Hearn, D.; Lewin, W. H. G.; Matilsky, T.; Mayer, W.; and Primini, F. "Discovery of a 283-second periodic variation in the X-ray source 3U 0900–40," *Astrophysical Journal*, Vol. 206 (1976), pp. L99–L102.

Michanowsky, George. *The Once and Future Star*. New York: Hawthorne Books, 1977.

Mitton, Simon. "Cygnus X-3: Dying swan or vigorous phoenix?" *New Scientist*, Vol. 56 (Oct. 26, 1972), pp. 200–2.

Miyamoto, S.; Fujii, M.; Matsuoka, M. Nishimura, J.; Oda, M.; Ogawara, Y.; and Ohta, S. "Measurement of the location of the X-ray source Cygnus X-1," *Astrophysical Journal*, Vol. 168 (1971), pp. L11–L14.

Rao, U. R.; Kasturirangan, K.; Sharma, D. P.; and Radha, M. S. "Observations of Cyg X-1 from Aryabhata, *Nature*, Vol. 260 (1976), pp. 307–8.

Sanford, P. W.; Ives, J. C.; Burnell, S. J. Bell; Mason, K. O.; and Murdin, P. "Ariel V and Copernicus measurements of the X-ray variability of Cyg X-1," *Nature*, Vol. 256 (1975), pp. 109–11.

Schreier, E.; Gursky, H.; Kellogg, E.; Tananbaum, H. and Giacconi, R. "Further observations of the pulsating X-ray source Cygnus X-1 from *Uhuru*," *Astrophysical Journal*, Vol. 170 (1971), pp. L21–L27.

Schreier, E.; Giacconi, R.; Gursky, H.; Kellogg, E.; and Tananbaum, H. "Discovery of the binary nature of SMC X-1 from *Uhuru*," *Astrophysical Journal*, Vol. 178 (1972), pp. L71–L75.

Shklovskii, I. S. "The nature of the X-ray source Sco X-1," *Soviet Astronomy—AJ*, Vol. 11, No. 5 (1968), pp. 749–55.

———. "On the nature of the source of X-ray emission of Sco XR-1," *Astrophysical Journal*, Vol. 148 (1966), pp. L1–L4.

Sommer, M.; Maurus, H.; and Urbach, R. "The hard X-ray spectrum of Cyg X-1 during the transition in November 1975," *Nature*, Vol. 263 (1976), pp. 752–53.

Tananbaum, H.; Kellogg, E.; Gursky, H.; Murray, S.; Schreier, E.; and Giacconi, R. "Measurement of the location of the X-ray sources Cygnus X-1 and Cygnus X-2 from *Uhuru*," *Astrophysical Journal*, Vol. 165 (1971), pp. L37–L41.

Tananbaum, H.; Gursky, H.; Kellogg, E. M.; Levinson, R.; Schreier, E.; and Giacconi, R. "Discovery of a periodic pulsating binary X-ray source in Hercules from *Uhuru*," *Astrophysical Journal*, Vol. 174 (1972), pp. L143–L149.

Thorne, Kip S. "The search for black holes," *Scientific American*, Vol. 231 (Dec. 1974), pp. 32–43.

Van de Heuvel, E. P. J., and Heise, J. "Centaurus X-3, possible reactivation of an old neutron star by mass exchange in a close binary," *Nature Physical Science*, Vol. 239 (1972), pp. 67–69.

Wade, C. M., and Hjellming, R. M. "Position and identification of the Cygnus X-1 radio source," *Nature*, Vol. 235 (1972), p. 271.

Webster, B. Louise, and Murdin, Paul. "Cygnus X-1—a spectroscopic binary with a heavy companion?" *Nature*, Vol. 235 (1972), pp. 37–38.

Wilson, A. M., and Carpenter, G. F. "X-ray outburst from Circinus X-1," *Nature*, Vol. 261 (1976), pp. 295–96.

Wright, Edward L.; Gottlieb, Elaine W.; and Liller, William. "Optical studies of *Uhuru* sources. XII. The light curve of Scorpius X-1 = V818 Scorpii, 1889–1974," *Astrophysical Journal*, Vol. 200 (1975), pp. 171–76.

Zeldovich, Y. B., and Novikov, I. D. "Relativistic Astrophysics. II. *Soviet Physics Uspekhi*, Vol. 7 (1965), pp. 763–88 (Part I), and Vol. 8 (1965), pp. 522–77 (Part II).

Zeldovich, Y. B., and Shakura, N. I. "X-ray emission accompanying the accretion of gas by a neutron star," *Soviet Astronomy—AJ*, Vol. 13, No. 2 (1969), pp. 175–83.

Chapter 11 · The Tests: Prediction and Observation

Bahcall, John N. "Masses of neutron stars and black holes in X-ray binaries," *Annual Review of Astronomy and Astrophysics*, Vol. 16 (1978), pp. 241–64.

Bahcall, J. N., Dyson, F. J.; and Katz, J. I. "Multiple star systems and X-ray sources," *Astrophysical Journal*, Vol. 189 (1974), pp. L17–L18.

Batten, A. H., and Olowin, R. P. "Black holes and binary stars," with a reply by S. W. Hawking and G. W. Gibbons, *Nature*, Vol. 234 (1971), pp. 341–42.

Brecher, K., and Morrison, P. "Rapidly rotating degenerate dwarfs as X-ray sources in binaries," *Astrophysical Journal*, Vol. 180 (1973), pp. L107–L112.

Brecher, K., and Caporaso, G. " 'Neutron' stars within the laws of physics," *Eighth Texas Symposium on Relativistic Astrophysics, Annals of the New York Academy of Sciences*, Vol. 302 (1977), pp. 471–81.

Bregman, Jesse; Butler, Dennis; Kemper, Edward; Koski, Alan; Kraft, R. P.; and Stone, R. P. S. "On the distance to Cygnus X-1 (HDE 226868)," *Astrophysical Journal*, Vol. 185 (1973), pp. L117–L120.

Cameron, A. G. W. "Evidence for a collapsar in the binary system E Aur," *Nature*, Vol. 229 (1971), pp. 178–80.

Chapline, George, and Nauenberg, Michael. "On the possible existence of quark stars," *Eighth Texas Symposium . . .* , op. cit., pp. 191–96.

Colgate, Stirling A. "Ejection of companion objects by supernovae," Nature, Vol. 225 (1970), pp. 247–48.

Connors, P. A., and Stark, R. F. "Observable gravitational effects on polarised radiation coming from near a black hole," *Nature*, Vol. 269 (1977), pp. 128–29.

Demarque, Pierre, and Morris, Stephen C. "Is there a black hole in E Aurigae?" *Nature*, Vol. 230 (1971), pp. 516–17.

DeWitt C., and DeWitt, B. S., eds. *Black Holes*. New York: Gordon & Breach, 1973.

Dolan, J. F.; Crannell, C. J.; Dennis, B. R.; Frost, K. J.; and Orwig, L. E. "Intensity transitions in Cyg XR-1 observed at high energies from OSO 8," *Nature*, Vol. 267 (1977) pp. 813–15.

Fabian, A. C.; Pringle, J. E.; and Whelan, J. A. J. "Is Cyg X-1 a neutron star?" *Nature*, Vol. 247 (1974), pp. 351–52.

Fechner, W. B.,; and Joss, P. C. "Quark stars with 'realistic' equations of state," *Nature*, Vol. 274 (1978), pp. 347–49.

Gibbons, G. W., and Hawking, S. W. "Evidence for black holes in binary star systems," *Nature*, Vol. 232 (1971), pp. 465–66.

Gursky, Herbert, and Ruffini, Remo, eds. *Neutron Stars, Black*

Holes and Binary X-ray Sources. Dordrecht: D. Reidel, 1975. Astrophysics and Space Science Library, Vol. 48.

Joss, P. C., and Rappaport, S. A. "Observational constraints on the masses of neutron stars," *Nature,* Vol. 264 (1976), pp. 219–22.

Leach, Robert W., and Ruffini, Remo. "On the masses of X-ray sources," *Astrophysical Journal,* Vol. 180 (1973), pp. L15–L18.

Margon, Bruce; Bowyer, Stuart; and Stone, Remington P. S. "On the distance to Cygnus X-1," *Astrophysical Journal,* Vol. 185 (1973), pp. L113–L116.

Metz, William D. "Astronomy from an X-ray satellite: Measuring the mass of a neutron star," *Science,* Vol. 179 (1973), pp. 884–85.

Penrose, Roger. "Gravitational collapse and space-time singularities," *Physical Review Letters,* Vol. 14 (1965), pp. 57–59.

Polidan, R. S.; Pollard, G. S. G.; Sanford, P. W.; and Locke, M. C. "X-ray emission from the companion to V861Sco," *Nature,* Vol. 275 (1978), pp. 296–97.

Pringle, J. E., and Rees, M. J. "Accretion disc models for compact X-ray sources," *Astronomy and Astrophysics,* Vol. 21 (1972), pp. 1–9.

Rappaport, S. A., and Joss, P. C. "The masses of neutron stars: Observational constraints," *Eighth Texas Symposium . . . ,* op. cit., pp. 460–70.

Rappaport, S.; Joss, P. C.; and McClintock, J. E. "The 3U 0900–40 binary system: Orbital elements and masses," *Astrophysical Journal,* Vol. 206 (1976), pp. L103–L106.

Rothschild, R. E.; Boldt, E. A.; Holt, S. S.; and Serlemitsos, P. J. "Millisecond temporal structure in Cygnus X-1," *Astrophysical Journal,* Vol. 189 (1974), pp. L13–L16.

———. "Submillisecond measurements of the low state of Cygnus X-1," ibid., Vol. 213 (1977), pp. 818–26.

Shakura, N. I., and Sunyaev, R. A. "Black holes in binary systems: Observational appearance," *Astronomy and Astrophysics,* Vol. 24 (1973), pp. 337–55.

———. "A theory of the instability of disk accretion on to black holes and the variability of binary X-ray sources, galactic nuclei and quasars," *Monthly Notices of the Royal Astronomical Society,* Vol. 175 (1976), pp. 613–32.

Thorne, Kip S., and Price, Richard H. "Cygnus X-1: An interpretation of the spectrum and its variability," *Astrophysical Journal,* Vol. 195 (1975), pp. L101–L105.

Trimble, Virginia; Rose, William K.; and Weber, Joseph. "A low-mass primary for Cygnus X-1?" *Monthly Notices of the Royal Astronomical Society,* Vol. 162 (1973), pp. 1p–3p.

Trimble, V. L., and Thorne, K. S. "Spectroscopic binaries and collapsed stars," *Astrophysical Journal,* Vol. 156 (1969), pp. 1013–19.

Walker, E. N. "HD 152667 and Sco X-2," *Monthly Notices of the Royal Astronomical Society,* Vol. 159 (1972), pp. 253–59.

Weisskopf, M. C., and Sutherland, P. G. "On the physical reality of the millisecond bursts in Cygnus X-1: Bursts and shot noise," *Astrophysical Journal,* Vol. 221 (1978), pp. 228–33.

Zeldovich, Y. B., and Guseynov, O. H. "Collapsed stars in binaries," *Astrophysical Journal,* Vol. 144 (1966), pp. 840–41.

Chapter 12 · White Holes, Wormholes, and Naked Singularities

Bardeen, Jame M. "Kerr metric black holes," *Nature,* Vol. 226 (1970), pp. 64–65.

Bardeen, J. M.; Carter, B.; and Hawking S. W. "The four laws of black-hole mechanics," *Communications in Mathematical Physics,* Vol. 31 (1973), pp. 161–70.

Carter, B. "Axisymmetric black hole has only two degrees of freedom," *Physical Review Letters,* Vol. 26 (1971) pp. 331–33.

Chandrasekhar, S. "The increasing role of general relativity in astronomy," Halley Lecture for 1972, op. cit.

Chapline, George F. "Cosmological effects of primordial black holes," *Nature,* Vol. 253 (1975), pp. 251–52.

Davies, P. C. W. "Uncensoring the universe," *The Sciences* (New York Academy of Sciences), Vol. 17 (1977), pp. 4–26 passim, excerpted from *Space and Time in the Modern Universe.* New York: Cambridge, University Press, 1977.

———. "Supertechnology," *New Scientist,* Vol. 77 (1978), pp. 787–88. Excerpted from his book *The Runaway Universe,* London: Dent, 1978.

Davies, P. C. W., and Taylor, J. G. "Do black holes really explode?" *Nature,* Vol. 250 (1974), pp. 37–38.

DeWitt C., and DeWitt, B. S., eds. *Black Holes.* New York: Gordon & Breach, 1973.

Einstein, A., and Rosen, N. "The particle problem in the general theory of relativity," *Physical Review,* Vol. 48 (1935), pp. 73–77.

Fuller, Robert W., and Wheeler, John A. "Causality and multiply-connected space-time," *Physical Review,* Vol. 128 (1962), pp. 919–23.

Gibbons, Garry, "Black holes are hot," *New Scientist,* Vol. 69 (Jan. 8, 1976), pp. 54–56.

Harrison B. Kent; Thorne, Kip S.; Wakano, Masami; and Wheeler, John Archibald. *Gravitation Theory and Gravitational Collapse.* Chicago, Ill.: University of Chicago Press, 1965, p. 137.

Hawking, S. W. "The occurrence of singularities in cosmology," *Proceedings of the Royal Society of London,* Series A, Vol. 294 (1966), pp. 511–21. See also Vol. 300 (1967), pp. 187–201.

————. "Black holes are white hot," *Seventh Texas Symposium on Relativistic Astrophysics, Annals of the New York Academy of Sciences,* Vol. 262 (1975), pp. 289–90.

————. "Black holes and unpredictability," *Eighth Texas Symposium . . . ,* op. cit., pp. 158–60.

————. "Particle creation by black holes," *Communications in Mathematical Physics,* Vol. 43 (1975), pp. 199–200.

————. "The quantum mechanics of black holes," *Scientific American,* Vol. 236 (Jan. 1977), pp. 34–40.

Hjellming, R. M. "Black and white holes," *Nature Physical Science,* Vol. 231 (1971), p. 20.

Hoyle, Fred. "Antimatter, galactic nuclei and theories of the universe," with a commentary by Gary Steigman, *Nature,* Vol. 224 (1969), pp. 477–81.

Israel, Werner. "Event horizons in static vacuum space-times," *Physical Review,* Vol. 164 (1967), pp. 1776–79.

Jeans, J. H. *Astronomy and Cosmogony.* New York: Cambridge University Press, 1928, P. 352.

Jelley, J. V.; Baird, G. A.; and O'Mongain, E. "Comments on the optical and radio detection of black hole explosions," *Nature,* Vol. 267 (1977), pp. 499–500.

Kaufmann, William J. III. *Relativity and Cosmology.* New York: Harper & Row, 1973, Chap. 6 and 7.

Lake, Kayll, and Roeder, R. C. "The present appearance of white holes," *Nature,* Vol. 273 (1978), pp. 449–50.

Meikle, W. P. S. "Upper limits for the radio pulse emission rate from exploding black holes," *Nature,* Vol. 269 (1977), pp. 41–42.

Misner, Charles W.; Thorne, Kip S.; and Wheeler, John Archibald. *Gravitation.* San Francisco Calif.: W. H. Freeman and Co., 1973.

Overbye, Dennis. "Out from under the cosmic censor: Stephen Hawking's black holes," *Sky and Telescope,* Vol. 54 (1977), pp. 84–108 passim.

Pathria, R. K. "The universe as a black hole," *Nature,* Vol. 240 (1972), pp. 298–99. See also comment by S. Chandrasekhar cited in *Science News,* Vol. 105 (1974), p. 99.

Penrose, Roger. "Black holes," *Scientific American,* Vol. 226 (May 1972), pp. 38–46.

———. "Black holes and gravitational theory," *Nature,* Vol. 236 (1972), pp. 377–80.

Penrose, Roger, and Floyd, R. M. "Extraction of rotational energy from a black hole," *Nature Physical Science,* Vol. 229 (1971), pp. 177–79.

Porter, N. A., and Weekes, T. C. "Optical pulses from primordial black hole explosions," *Nature,* Vol. 267 (1977), pp. 500–1.

Press, William H., and Teukolsky, Saul A. "Floating orbits, superradiant scattering and the black-hole bomb," *Nature,* Vol. 238 (1972), pp. 211–12.

———. "Perturbations of a rotating black hole. II. Dynamical stability of the Kerr metric," *Astrophysical Journal,* Vol. 185 (1973), pp. 649–73.

Rees, Martin; Ruffini, Remo; and Wheeler, John Archibald. *Black Holes, Gravitational Waves and Cosmology.* New York: Gordon & Breach, 1975, pp. 51, 286–307.

Robinson, D. C. "Uniqueness of the Kerr black hole," *Physical Review Letters,* Vol. 34 (1975), pp. 905–06.

Ruffini, Remo, and Wheeler, John A. "Introducing the black hole," *Physics Today,* Vol. 24 (1971), pp. 30–41.

Simpson, Michael, and Penrose, Roger. "Internal instability in a Reissner-Nordström black hole," *International Journal of Theoretical Physics,* Vol. 7 (1973), pp. 183–97.

Thorne, Kip S. "Gravitational Collapse," *Scientific American,* Vol. 217 (Nov. 1967), pp. 88–98.

Weekes, T. C., and Porter, N. A. "A search for X-ray bursts from the explosive evaporation of black holes," *Proceedings of the Fifteenth International Conference on Cosmic Rays,* Plovdiv, Bulgaria (Aug. 1977) (in press).

Wood, Lowell; Weaver, Thomas; and Nuckolls, John. "New approaches to CTR: General relativistic power plants," *Annals of the New York Academy of Sciences,* Vol. 251 (1975), pp. 623–31.

Chapter 13 · The Universe—Open, Shut, or According to Hoyle

Alfvén, Hannes. "Antimatter and the development of the metagalaxy," *Reviews of Modern Physics,* Vol. 37 (1965), pp. 652–65.

———. "Plasma physics applied to cosmology," *Physics Today,* Vol. 24 (Feb. 1971), pp. 28–33.

Bolyai, Farkas, quoted in "Geometry, non-Euclidean," by H. S. MacDonald Coxeter, Encyclopaedia Britannica (Macropaedia), Vol. 7 (1976), p. 1113.

Burbidge, G. "Was there really a big bang?" *Nature,* Vol. 233 (1971), pp. 36–40.

Burbidge, E. Margaret; Burbidge, G. R.; Fowler, William A.; and Hoyle, F. "Synthesis of the elements in stars," *Reviews of Modern Physics,* Vol. 29 (1957), pp. 547–650.

Burbidge, G. R., and Hoyle, F. "Matter and anti-matter," *Il Nuovo Cimento,* Vol. 4 (1956), pp. 558–64.

Davies, P. C. W. "A new theory of the universe," *Nature,* Vol. 255 (1975), pp. 191–22.

Dicke, R. H., "Gravitation and the Universe," *Science Journal,* Vol. 2 (Oct. 1966), pp. 95–100.

Dicke, R. H.; Peebles, P. J. E.; Roll, P. G.; and Wilkinson, D. T. "Cosmic black-body radiation," *Astrophysical Journal,* Vol. 142 (1965), pp. 414–19.

Einstein, A., in *The Principle of Relativity,* H. A. Lorentz, et al. London: Methuen, 1923. Quoted by Tinsley, *Physics Today,* Vol. 30 (June 1977), p. 34.

———. *The Meaning of Relativity,* 5th ed. Princeton, N.J.: Princeton University Press, 1955, pp. 107–8.

Hoyle, F. "A new model for the expanding universe," *Monthly Notices of the Royal Astronomical Society,* Vol. 108 (1948), pp. 372–82.

———. "Recent developments in cosmology," *Nature,* Vol. 208 (1965), pp. 111–14.

———. "On the origin of the microwave background," *Astrophysical Journal,* Vol. 196 (1975), pp. 661–70.

———. "The origin of the universe," Frank Nelson Doubleday Lecture, Washington, D.C.: National Museum of History and Technology, 1975.

Hoyle, F., and Narlikar, J. V. "On the avoidance of singularities in

C-field cosmology," *Proceedings of the Royal Society of London,* Series A, Vol. 278 (1964), pp. 465–78.

————. "On the effects of nonconservation of baryons in cosmology," *Proceedings of the Royal Society of London,* Series A, Vol. 290 (1966), pp. 143–61. See also the following papers.

Narlikar, J. V. "Singularity and matter creation in cosmological models," *Nature Physical Science,* Vol. 242 (1973), pp. 135–36.

Thomsen, Dietrick E. "Cosmology according to Hoyle," *Science News,* Vol. 107 (June 14, 1975), pp. 386–87.

Wheeler, John Archibald. "Our universe: The known and the unknown," *The American Scholar,* Vol. 37 (Spring 1968), pp. 248–74.

Chapter 14 · The Arrow of Time

Bekenstein, Jacob D. "Black holes and entropy," *Physical Review D—Particles and Fields,* Vol. 7 (1973), pp. 2333–46.

Davies, P. C. W. "Closed time as an explanation of the black body background radiation," *Nature Physical Science,* Vol. 240 (1972), pp. 3–5. See also commentary in *Nature,* Vol. 240 (1972), p. 15.

Eddington, A. S. *The Nature of the Physical World,* New York: Macmillan, 1933, pp. 63–64, 71–73.

Gal-Or, Benjamin. "The crisis about the origin of irreversibility and time anisotrophy," *Science,* Vol. 176 (1972), pp. 11–17.

Gold, T. "The arrow of time," *American Journal of Physics,* Vol. 30 (1962), pp. 403–10.

Peat, F. D. "Black holes and temporal ordering," *Nature,* Vol. 239 (1972), p. 387.

Sachs, Robert G. "Time reversal," *Science,* Vol. 176 (1972), pp. 587–97.

Thomsen, Dietrick E. "A cosmological triple play—Is ours only one of three universes?" *Science News,* Vol. 105 (Feb. 16, 1974), p. 109.

Tipler, Frank J. "Black holes in closed universes," *Nature,* Vol. 270 (1977), pp. 500–1.

Wheeler, J. A. "At last a sane look at the 'arrow of time,' " *Physics Today,* Vol. 28 (June 1975), pp. 49–50.

Chapter 15 · The "State of the Universe Message"

Baldwin, Jack A. "Luminosity indicators in the spectra of quasi-stellar objects," *Astrophysical Journal,* Vol. 214 (1977), pp. 679–84.

Baldwin, Jack A.; Burke, William L.; Gaskell, C. Martin; and Wampler, E. J. "Relative quasar luminosities determined from emission line strengths," *Nature,* Vol. 273 (1978), pp. 431–35.

Barrow, John D., and Tipler, Frank J. "Eternity is unstable," *Nature,* Vol. 276 (1978), pp. 453–59.

Davidsen, Arthur F.; Hartig, George F.; and Fastie, William G. "Ultraviolet spectrum of quasi-stellar object 3C 273," *Nature,* Vol. 269 (1977), pp. 203–6.

Eddington, Sir Arthur Stanley. *The Nature of the Physical World.* New York: Macmillan, 1933, p. 86.

Fang Li-zhi; Zhou You-yuan; Cheng Fu-zhen; and Chu Yao-quan, paper in *Acta Astronomica, Sinica,* Vol. 17, No. 2 (1977), p. 134, trans. in *Chinese Astronomy,* Vol. 1, No. 2 (in press). Cited by T. Kiang in "More evidence for a closed universe from QSO's," *Nature,* Vol. 270 (1977), pp. 205–6.

Faulkner, John. "G-wizardry at Dallas," *Nature,* Vol. 253 (1975), pp. 231–32.

Gott, J. Richard III; Gunn, James E.; Schramm David N.; and Tinsley, Beatrice M. "An unbound universe?" *Astrophysical Journal, Vol. 194* (1974), pp. 543–53.

Kristian, Jerome; Sandage, Allan; and Westphal, James A. "The extension of the Hubble diagram. II. New red shifts and photometry of very distant galaxy clusters: First indication of a deviation of the Hubble diagram from a straight line," *Astrophysical Journal,* Vol. 221 (1978), pp. 383–94.

Pagel, B. E. J. "Ultraviolet observation of a quasar spectrum," *Nature,* Vol. 269 (1977), pp. 195–96.

Penzias, A. A. "The riddle of cosmic deuterium," *American Scientist,* Vol. 66 (1978), pp. 291–97.

———. "Interstellar HCN, HCO+, and the galactic deuterium gradient," *Astrophysical Journal,* Vol. 228 (1979), pp. 430–34.

Rogerson, John B., Jr., and York, Donald G. "Interstellar deuterium abundance in the direction of Beta Centauri," *Astrophysical Journal,* Vol. 186 (1973), pp. L95–L98.

Sandage, Allan. "The light travel time and the evolutionary correction to magnitudes of distant galaxies," *Astrophysical Journal,* Vol. 134 (1961), pp. 916–26.

————. "The existence of a major new constituent of the universe: The quasi-stellar galaxies," *Astrophysical Journal,* Vol. 141 (1965), pp. 1560–78.

————. "Cosmology: A search for two numbers," *Physics Today,* Vol. 23 (Feb. 1970), pp. 34–41.

————. "The redshift-distance relation. VI. The Hubble diagram from S20 photometry for rich clusters and sparse groups: A study of residuals," *Astrophysical Journal,* Vol. 183 (1973), pp. 731–42.

Kazanas, Demosthenes; Schramm, David N.; and Hainebach, Kem. "A consistent age for the universe," *Nature,* Vol. 274 (1978), pp. 672–73.

Tayler, R. J. "Neutrino stability and cosmological helium production," *Nature,* Vol. 274 (1978), pp. 232–33.

Tinsley, Beatrice M. "The cosmological constant and cosmological change," *Physics Today,* Vol. 30 (June 1977), pp. 32–38.

————. "Accelerating universe revisited," *Nature,* Vol. 273 (1978), pp. 208–10.

Chapter 16 · Black Holes—Everywhere or Nowhere

Bahcall, J. N., and Ostriker, J. P. "X-ray pulses from globular clusters," *Nature,* Vol. 262 (1976), pp. 37–38.

Brecher, Kenneth. "Cosmic X-ray burst," *Nature,* Vol. 261 (1976), p. 542.

Clark, G. W.; Jernigan, J. G.; Bradt, H.; Canizares, C.; Lewin, W. H. G.; Li, F. K.; Mayer, W.; and McClintock, J. "Recurrent brief X-ray bursts from the globular cluster NGC 6624," *Astrophysical Journal,* Vol. 207 (1976), pp. L105–L108.

Cruddace, R. G.; Fritz, G.; Shulman, S.; and Friedman, H. "High-resolution observations of X-ray sources at the galactic center," *Astrophysical Journal,* Vol. 222 (1978), pp. L95–L98.

Evans, W. D.; Belian, R. D.; and Connor, J. P. "Observations of intense cosmic X-ray bursts," *Astrophysical Journal,* Vol. 207 (1976), pp. L91–L94.

Fabian, A. C.; Maccagni, D.; Res. M.J.; and Stoeger, W. R. "The nucleus of Centaurus A," *Nature,* Vol. 260 (1976), pp. 683–85. See also J. H. Beall et al., "Radio and X-ray variability of the nucleus of Centaurus A (NGC 5128)," *Astrophysical Journal,* Vol. 219 (1978), pp. 836–44.

Field, George B., and Perrenod, Stephen C. "Constraints on a dense hot intergalactic medium," *Astrophysical Journal,* Vol. 215 (1977), pp. 717–22.

Forman, W., and Jones, C. "*Uhuru* observations of an X-ray burst at high galactic latitude centered on the X-ray globular cluster NGC 1851," *Astrophysical Journal,* Vol. 207 (1976), pp. L177–L180.

Forman, William; Jones, Christine; Cominsky, Lynn; Julien, Paul; Murray, Stephen; Peters, Geraldine; Tannanbaum, Harvey; and Giacconi, Riccardo. "Fourth *Uhuru* catalogue of X-ray sources," *Astrophysical Journal* (Supplement), Vol. 38, No. 4 (1978).

Friedman, Herbert, quoted in "More and more mass: Universe is closing," *Science News,* Vol. 114, No. 12 (1978), p. 198.

Green, Louis C. "Some new developments in X-ray astronomy," *Sky and Telescope,* Vol. 53 (May 1977), pp. 340–43.

Grindlay, Jonathan E. "Diffuse γ-ray background from Seyfert galaxies," *Nature,* Vol. 273 (1978), p. 211.

Grindlay, J., and Grusky, H. "Scattering model for X-ray bursts: Massive black holes in globular clusters," *Astrophysical Journal,* Vol. 205 (1976), pp. L131–L133.

Grindlay, J.; Gursky, H.; Schnopper, H.; Parisgnault, D. R.; Heise, J.; Brinkman, A. C.; and Schrijver, J. "Discovery of intense X-ray bursts from the globular cluster NGC 6624," *Astrophysical Journal,* Vol. 205 (1976), pp. L127–L130.

Grindlay, J. E., and Liller, William. "Evidence for ionized hydrogen in the cores of globular clusters," *Astrophysical Journal,* Vol. 216 (1977), pp. L105–L109.

Gull, S. F., and Northover, K. J. E. "Detection of hot gas in clusters of galaxies by observation of the microwave background radiation," *Nature,* Vol. 263 (1976), pp. 572–73.

Gursky, Herbert, "Nature of X-ray bursters," *Eighth Texas Symposium* . . . , op. cit., pp. 197–209.

Harrison, E. R. "Has the sun a companion star?" *Nature,* Vol. 270 (1977), pp. 324–26.

Henrichs, H. F., and Staller, R. F. A. "Has the sun really got a companion star?" *Nature,* Vol. 273 (1978), pp. 132–34.

Hughes, V. A., and Viner, M. R. "W_3(OH): a 'runaway' compact object?" *Astrophysical Journal,* Vol. 222 (1978), pp. L27–L31.

Joss, P. C., and Rappaport, S. "A simple physical model for X-ray burst sources," *Nature,* Vol. 265 (1976), pp. 222–24.

Kellermann, K. I.; Shaffer, D. B.; and Clark, B. G. "The small radio source at the galactic center," *Astrophysical Journal*, Vol. 214 (1977), pp. L61–L62.

Lewin, W. H. G.; Doty, J.; Clark, G. W.; Rappaport, S. A.; Bradt, H. V. D.; Doxsey, R.; Hearn, D. R.; Hoffman, J. A.; Jernigan, J. G.; Li, F. K.; Mayer, W.; McClintock, J.; Primini, F.; and Richardson, J. "The discovery of rapidly repetitive X-ray bursts from a new source in Scorpius," *Astrophysical Journal*, Vol. 207 (1976), pp. L95–L99.

Lewin, W. H. G.; Hoffman, J. A.; Doty, J.; Hearn, D. R.; Clark, G. W.; Jernigan, J. G.; Li, F.K.; McClintock, J. E.; and Richardson, J. "Discovery of X-ray bursts from several sources near the galactic centre," *Monthly Notices of the Royal Astronomical Society*, Vol. 177 (1976), pp. 83p–92p.

Lewin, Walter H. G. "X-ray burst sources," *Monthly Notices of the Royal Astronomical Society*, Vol. 179 (1977), pp. 43–53.

Lewin, Walter H. G., and Joss, Paul C. "X-ray burst sources," *Nature*, Vol. 270 (1977), pp. 211–16. (References constitute a fairly complete bibliography.)

Lewin, W. H. G.; Hoffman, J. A.; Doty, J.; Clark, G. W.; Swank, J. H.; Becker, R. H.; Pravdo, S. H.; and Serlemitsos, P. J. "Galactic distribution of X-ray burst sources," *Nature*, Vol. 267 (1977), pp. 28–30.

Liller, M. H., and Liller, W. "Preliminary photometry of the X-ray globular cluster NGC 6624," *Astrophysical Journal*, Vol. 207 (1976), pp. L109–L111.

Liller, William, "Optical Observations of X-ray globular clusters," *Annals of the New York Academy of Sciences*, Vol. 302 (1977), pp. 248–60.

————. "Searches for the optical counterparts of the X-ray burst sources MXB 1728–34 and MXB 1730–33," *Astrophysical Journal*, Vol. 213 (1977), pp. L21–L24.

Lynden-Bell, D., in *Nuclei of Galaxies*, ed. D. J. K. O'Connell, S. J. (Vatican: April 13–18, 1970), pp. 527–38.

Murray, S. S.; Forman, W.; Jones, C.; and Giacconi, R. "Evidence for X-ray emission from superclusters of galaxies determined from *Uhuru*," *Astrophysical Journal*, Vol. 219 (1978), pp. L89–L93.

Oke, J. B., and Sargent, W. L. W. "Seyfert galaxies," *Science Journal*, Vol. 6 (Feb. 1970), pp. 57–61.

Oort, J. H. "The galactic center," *Annual Review of Astronomy and Astrophysics*, Vol. 15 (1977), pp. 295–362.

Ostriker, Jeremiah P.; Spitzer, Lyman Jr.; and Chevalier, Roger A. "On the evolution of globular clusters," *Astrophysical Journal*, Vol. 176 (1972), pp.L51–L56.

Peebles, P. J. E. "Dead galaxies?" *Astrophysical Journal*, Vol. 154 (1968), pp. L121–L123.

———. "Star distribution near a collapsed object," *Astrophysical Journal*, Vol. 178 (1972), pp. 371–75.

Pounds, K. A. "The Ariel V high latitude catalogue," *Annals of the New York Academy of Sciences*, Vol. 302 (1977), pp. 361–85.

Pounds, Ken. "Rise and fall of an X-ray star," *New Scientist*, Vol. 69 (Mar. 4, 1976), pp. 494–96.

Press, William H., and Gunn, James E. "Method for detecting a cosmological density of condensed objects," *Astrophysical Journal*, Vol. 185 (1973), pp. 397–412.

Readhead, A. C. S.; Cohen, M. H.; and Blandford, R. D. "A jet in the nucleus of NGC6251," *Nature*, Vol. 272 (1978), pp. 131–34.

Rees, Martin J. "Quasar theories," *Eighth Texas Symposium . . .* , op. cit., pp.613–35.

Sargent, W. L. W.; Young, Peter J.; Boksenberg, A.; Shortridge, Keith; Lynds, C. R.; and Hartwick, F. D. A. "Dynamical evidence for a central mass concentration in the galaxy M87," *Astrophysical Journal*, Vol. 221 (1978), pp. 731–44.

Seward, F. D., and Liller, William. "Was the bright transient X-ray source Centaurus XR-4 a globular cluster?" *Publications of the Astronomical Society of the Pacific*, Vol. 89 (1977), pp. 696–98. (On NGC 5824)

Shapiro, Stuart L., and Elliot, James L. "Are BL Lac-type objects nearby black holes?" *Nature*, Vol. 250 (1974), pp. 111–13.

Tarter, Jill C. "Some problems with the interpretation of recent microwave background observations in the direction of galaxy clusters, or, beware of negative antenna temperatures," *Astrophysical Journal*, Vol. 220 (1978), pp. 749–55.

Wheeler, J. C., and Shields, G. A. "Origin of the black hole in Cyg X-1," *Nature*, Vol. 259 (1976), pp. 642–43.

Waggett, P. C.; Warner, P. J.; and Baldwin, J. E. "NGC6251, a very large radio galaxy with an exceptional jet," *Monthly Notices of the Royal Astronomical Society*, Vol. 181 (1977), pp. 465–74.

Wolfendale, Arnold W. "Origin of the diffuse cosmic gamma rays," *Nature*, Vol. 274 (1978), pp. 314–15.

Wolfendale, A. W., and Worrall, Diana M. "Cosmic rays near the galactic centre," *Nature*, Vol. 263 (1976), pp. 482–83.

Wyller, Arne A. "Observational aspects of black holes in globular clusters," *Astrophysical Journal*, Vol. 160 (1970), pp. 443–49.

Young, Peter J.; Westphal, James A.; Kristian, Jerome; Wilson, Christopher P.; and Landauer, Frederick P. "Evidence for a supermassive object in the nucleus of the galaxy M87 from SIT and CCD area photometry," *Astrophysical Journal*, Vol. 221 (1978), pp. 721–30.

Zeldovich, Y. B., and Novikov, I. D. "The hypothesis of cores retarded during expansion and the hot cosmological model," *Soviet Astronomy—AJ*, Vol. 10 (1967), pp. 602–3.

Index

stick (*see* Red shift). *See also under* name

Astronomy, 249; distance measuring, 101–02, 103 *ill.*, 104–05; modern, 123; observational devices, 93; *ill.*, 105, 298; radio, 91 (*see* Radio astronomy); red shift, 103 *ill.*, 104; traditions, 143–44; X-ray, 169–70, 200–01

Astrophysical Journal, 122, 198; editor, 54, 122

Astrophysicists, 107, 113, 172, 207, 226, 304–05

Astrophysics, 199, 247–48; puzzles, 292. *See* Harvard-Smithsonian Center *and* Texas Conferences

Asymmetrical collapse, 230–32

Atluri, C.R., 31, 34, 36, 40

Atmosphere: "light" absorption, 163; ionization, 167; shock waves, 14

Atomic Age, 49–50, 253–54

Atomic bomb, 26, 34–36, 47, 86, 191, 193; chain reaction, 253; "father," 79; first, 49, 253–54; Los Alamos project, 85, 86; mass-energy relationship, 63; Tunguska explosion, 27–27. *See* Oppenheimer, J. Robert

Atomic clock, 73, 144, 295; relativity, 73–75

Atomic energy, 47. *See* Nuclear energy

Atomic excitation, 160–61

Atomic particles, 227

Atomic power plants, 27

Atoms, 14, 32–34, 47–50, 61, 278–79; ionized, 118; nature of, 48–50; quantum behavior, 48; solar system analogy, 48

Australia, 94, 99, 118–19, 120, 160, 171, 176; Anglo-Australian Observatory, 160; Commonwealth Scientific and Industrial Research Organization, 4; Woomera range, 187

Austria, 48, 249; *Anschluss*, 249

Baade, Walter, 60–62, 99–101, 105, 107, 112, 145, 151, 159 *ill.*, 161, 171

"Baade's star," 151–52, 153, 156–57 *ill.*, 158

Bahcall, John N., 197, 290

Bahcall, Neta, 197–98

Baldwin, Jack A., 275

Balloons, 159, 168, 176, 290

Balmer series, 119

Bardeen, James M., 231, 294

Barometric pressure: recording, 14

Bartol Research Foundation, 290

Baryons, 246, 250

Bekenstein, Jacob D., 263

Belgium, 202, 248

Bell Telephone Laboratories, 91, 132, 136; telescope, 283

Ben-Menahem, Ari, 30

Bergmann, Peter G., 123, 127

Bessel, Friedrich W., 43, 44

Betelgeuse (star), 80

Bethe, Hans, 253

Big Bang, 38, 61, 104, 123, 125, 234, 235, 238, 241, 244, 247, 266, 274, 277, 278, 279; concept, 244; cosmology, 246; energy, 113; evidence of, 136; fireball, 132, 133 *ill.*, 135, 285 (*see* Fireball); "glow," 280, 283, 287, 288–90; Hoyle, 249–51 (*see under* name); new, 274; theory, 61, 135, 136; time of, 269, 276

Big Dipper, 43

Binary systems, 189, 193

Birbeck College (London), 228

Black Cloud, The, 113

Black dwarfs, 303

Black holes, 26, 41, 92, 126–27, 206ff., 213ff., 225ff., 232ff., 251, 267, 280, 292, 297, 299–300, 303; "absurdity," 86; alternative, 215 (*see* Quarks); "bomb," 237; "bursters," 293; candidate, 188, 193, 202, 208, 218, 292 (*see* Cyg X-1); collapse, 289; "color photograph,"

216; companion star, 232; concept, 32, 49–50, 65, 67; birth of, 79, theoretical core, 125(*see* Schwarzschild radius); "conventional," 38, 294; core, 46; density, 296–97; description, 37–38, 85; detection, 304; as energy source, 229–30, 237–38, 250–51, 295; entropy, 262–63; ergosphere, 232; everywhere/nowhere? references, 328–32; evidence, 161, 168, 216, 219ff., 223; evolution, 215; existence, 13, 303–05; universe, 243ff., 285; explanations, 126; formation, 252–53; giant, 286; idea born: references, 310–13; impact, 13; infinities, 304; light emission, 304; location, 285–300; "memory," 239–41; "mini," 37, 38, 232–34, 236, 238, 285, 293; numbers of, 285ff.; Oppenheimer, 85–87, 124; overfeeding, 294; panacea, 293; possibility of, 230; pre-conditions for, 304; proponents, 213ff., 221, 238; and relativity, 211; Schwarzschild radius, 80; "seed," 285; "singularity," 13; sizes, 235; skeptics, 211ff., 221, 292–93, 295; spin rate, 231, 232–33; supermassive, 290, 294, 296–99 (*see also* M 87); term, 126, 224; theorists, 38–40, 211, 227–28, 231ff.; Tunguska, 36ff., *See also* "Event horizon" *and* Hawking, Stephen

BL Lacertae, 130, 295, 296
BL Lacs, 130
Bloch, Felix, 52–53 *ill.*
Bohr, Niels, 47, 52–53 *ill.*; Institute, 52–53
Bolton, C.T. (Tom), 206–07
Bolton, John G., 94
Bolyai, Farkas, 244
Bolyai, János, 244
Bondi, Hermann, 146, 248–49, 263
Boötes, 103–05

Boynton, P.E. 195, 198
Braes, L.L.E., 202
Brahe, Tycho, 59
Bratislava, Astronomical Institute of the Slovak Academy of Sciences in, 40–41
Brazil, 69
Brecher, Kenneth, 195, 214, 215–16, 292–93
Brecher-Morrison black-hole proposal, 216
Bremsstrahlung ("braking radiation"), 289
Broglio, Luigi, 177–79, 181
Brown, John C., 40
Bulgaria, 60
Burbidge, Geoffrey, 113, 125, 129, 219, 247, 295; Hoyle, 249–50, 251
Burbidge, Margaret, 113, 114 *ill.*, 125, 249, 251
Burnell, Jocelyn Bell, 139–43, 141 *ill.*, 145, 161
"Bursters," 291–92; 1728–34, 293
Byurakan Observatory, 106

California, 67,91,187, 213,235, 248, 270
Cal Tech (California Institute of Technology), 60, 85, 113, 117, 208, 216, 220, 231, 278, 286, 294, 295; A catalogue, 113; dishes, 117; Owens Valley Radio Observatory, 295; Palomar group, 117, 120
Cambridge, Mass., 158, 168, 197
Cambridge Catalogue of Radio Sources, 105, 117
Cambridge University, 38, 69, 105, 129, 145, 175, 202, 231, 248, 249, 288, 295; astronomy group, 113, 145, 161; graduate students, 140; Institute of Theoretical Astronomy, 200, 249; interferometer, 99; Optical Observatory, 144–45
Cameron, Alastair G.W., 145, 218
Canada, 85, 199, 206, 231, 249,

[336]

sources, 117, 127; (see Radio galaxies); red shift relationship, 104–05; spectra, 103 ill., 104; spiral, 106–07, 110–11, ill., 226; variability, 118. See also under name and NGC 5128

Galileo 115

Gamma rays, 33; 159, 212, 233–35; glow 247, 252–53, 287; origin, 212

Gamow, George, 52–53 ill., 61, 135–36, 247

Gantries, 167

Gas (hot), 287–88

Gauss, Carl Friedrich, 244

Geiger counters, 169

Gemini spacecraft, 175

General Electric Research Laboratory, 107

Geneva, Switzerland, 214

Gentry, Robert V., 36

Geometry, non-Euclidean, 244

"Geon," 126

George Washington University, 61

Germany, 46, 60, 67, 68, 79, 99, 127, 222, 244, 257, 263, 293, 295; Austria, 249; V-2 rockets, 163–64; World War II, 92

Giacconi, Riccardo, 168ff., 175, 176, 178, 187, 189, 207, 209, 288

Ginzburg, Vitali, 112, 124

"Glitches," 160

Globular clusters, 290–93; NGC 1851, 291; NGC 6624, 291

God, 248

Goddard Space Flight Center. See under NASA

Gold, Thomas, 146–51, 161, 172, 248, 263, 265; reputation, 146

Goozh, Paul, 179

Gravitation: astronomer's concept, 70; textbook, 230, 237

Gravitational collapse, 89, 101, 113, 125, 253, 294, 304; Dallas symposium, 124; in reverse, 243. See also Fowler; Hoyle, and Stars

Gravitational energy, 150 ill., "Gravitational radius," 80

"Gravitational red shift," 67, 74

Gravitational waves, 221–22

Gravity, 37, 43, 44, 65ff., 191; acceleration equivalence, 65–66 antigravity, 125; black holes, 37–38; earth's, 37–38; Einstein theory, 85–86 (see Einstein, Albert); energy, 113, 115, 304; force, 37, 65; Laplace, 81, 83–85; light, 67–68, 87; Newton theory, 87–88; "relativistic" treatment, 62–63; simulation (in spacecraft), 72; space, 65, 76; time, 65; variations, 44

Green Bank, W. Va., 202, 204, 205

Greenstein, Jesse L., 117, 120

Grindley, Jonathan E., 291

Groves, Gen. Leslie R., 254

"Guest star," 57–59

Gulf of Guinea, 69

Gull, S. F., 288

Gum, Colin S., 160

Gum Nebula, 160

Gunn, James 279–80, 281, 286, 295

Gursky, Herbert, 170, 205

Guseynov, 216, 218

GX 339-4, 221

Haber, Floyd, 107

Hafele, Joseph C., 72–74

Hale, George Ellery, 67

Hale Observatories, 93 ill., 103 ill., 110 ill., 270, 274, 278

"Half life," 35

Halley's Comet, 59

Harbin, Manchuria, 85

Harrison, Edward, 285, 305

Hartig, George, F., 273 ill., 277

Harvard University, 49, 71, 85, 128, 196, 292; black holes, 213; Jefferson Physical Laboratory, 73; Observatory, 175, 196–98, 205; Smithsonian Center for Astrophysics, 199, 288, 289. 291

Quantum behavior, 48, 112–13
Quantum mechanics, 305
Quantum theory, 227, 232, 240
"Quarks," 32, 41, 215, 305; "bag star,"215
Quasars, 117–37, 189, 193, 251–52, 273 *ill.*, 274, 275–78, 286–87, 294, 295, 297, 305; brightest, 122, 127–28; brilliance, 225, 300; discovery, 172–73, 277–78; distance, 122–23, 128, 129, 136; energy, 122, 292–93; source, 125, 128, 296; explanation, 296; first, locating, 171 (*see* Hazard); "geons," 126; identifying, 139; name, 120–22; nature, 128–30, number found, 136–37, 289; red shifts, 128–30; references, 311–12; scintillation, 139–40; as "standard candles," 282–83; 3C 279, 294
"Quasistellar radio sources," 122
Quito, Ecuador, 178, 181

Radar, 142, 249; antiaircraft, 92, 94; jamming, 92
Radiation: electromagnetic, 74, 126; nature, 265; supernovas, 59; synchrotron, 112
Radiation belts (Van Allen), 167
Radioactivity: forms, 35; Tunguska, 27, 29
Radio astronomy (ers), 91–92, 94, 105, 106, 142, 144, 147, 159, 163, 201–02, 279, 295, 300, 305; antennae, 133 *ill.*, 132, 147, 148–49 *ill.*, 171–72; birth, 91, 140; interferometry, 94, 99; new instruments, 91, 98–99; pioneers, 94; techniques, 94
Radio blackouts, 167–68
Radio communications, past horizon, 164
Radio emissions, 92, 94, 130, 132
Radio galaxies, 127–29, 294–95; energy source, 294, 297, 300; mapping, 295

Radio interference, sources, 117
Radio pulsars, 193, 215, 304
"Radio sky," 94
Radio sources, 105–07, 117, 118, 119, 123, 299–300; between civilizations, 124, 131; electrical interference, 142; formation, 123; mapping, 160; "radio galaxies," 106; scintillation or "twinkling," 139. *See* Cal Tech A catalogue
Radio telescopes. *See under* Telescopes
Radio waves, 67, 94, 169, 200–01; sources, 94, 101, 104
Rainbow, 104
Raman, Chandrasekhara V., 50, 54
Ramanathan, K. R., 200
"Raman effect," 50
Reasoning: *reductio ad absurdum,* 54
Reber, Grote, 92
Rebka, Glen A., Jr., 73
Red Sea, 50
Red shift, 103 *ill.*, 104, 105–06, 120–22, 127, 250, 267, 270, 273 *ill.*; gravitational, 120. *See* Quasars
Rees, Martin, 129, 220, 240, 296–97, 300
Reeves, Hubert, 279
Reifenstein, Edward C., III, 147, 151
Relativity, 84, 227; black holes 304; effects, 71–72; general theory, 63, 71, 73–74, 79, 211, 226, 227, 240, 265; authorities, 127; black holes, 211; confirmation of, 221; effects, 73; implication, 126; principle, 227; origins, 71; predictions, 304; "special" theory, 63, 71, 73, 89; theorists, 122ff.; time-warping effects, 76. *See* Einstein, Albert
Research, 159, 236
Rice, Oscar K., 52–53 *ill.*
Riemann, George Friedrich B., 244
Robinson, David C., 235
Robinson, Ivor, 123

Laboratory, 170–73, 193, 200, 221, 288, 293. *See also* NASA *and* National Radio Astronomy Observatory
Universe, 14, 243–54, 265, 269–83, 286, 287–90; age, 254; assumptions about, 123; "baryon number," 247; birth, 37, 61, 105, 238–39, 240–41, 246–49, 250–51 (*see* Big Bang); black holes, 285; "bounce," 252–53, 305; closed, 278ff., 286, 288, 303, 325–26; collapse, 225, 239–41, 244–45, 251–54, 267, 303–05; concept of, change in, 69–70; density, 244, 281; "edge," 137, 144; expansion, 101, 104, 119–20, 123, 244ff., 251–52, 264–65, 269ff., 283, 285; Einstein, 265; entropy, 263; evidence, 250–51; rate, 277ff.; slowing (q_0), 122, 269–70, 271–75, 273 *ill.*, 280; fate, 243ff., 257ff., 269, 270ff., 277 (see Time, arrow of); fireball residue, 150; glow, 287–88; Hoyle, 125, 243, 246–48; life on, 267; missing mass, 286, 288; nature of, 104–05, 244ff., 251–52, 263; open, 243ff., 277ff., 303, 325–26; questions about, 99, 245, 269ff., 305; sky (darkness), 263–64; slowing (*see under* expansion): "State of, Message," 277ff., 281–82, 327–28; "steady state," 113, 125, 146, 243, 246, 248–51 (*see* Hoyle); symmetry, 34; "temperature," 280; test, 269. *See also* Big Bang
Universities: Alberta, 231; Arizona, 158; Steward Observatory, 152, 158; British Columbia, 86; California, 236; Berkeley, 85, 107, 197, 290; Los Angeles, 31; Chicago, 49, 278; Glasgow, 40; Leicester, 187, 287; Leningrad, 61; London, 298; Louvain, 248; Maryland, 76, 213, 222; Massa-

chusetts, 222, 285; Michigan, 296; Rochester, 196–97; Mees Observatory, 196; Sheffield, 40; Texas, 122–23; Center for Relativity Theory, 37; Tokyo, 173, 175, 205; Observatory, 175–76; Toronto, 206; David Dunlap Observatory, 206; Washington, 195, 198, 231, 294 *See also under* name
Uranium, 27, 189
Ursa Major (Great Bear), 103 *ill.*

V-2 rockets, 163–64, 167; launch (1946), 164, 165 *ill.*, 167
Van Allen, James A., 167
Van Allen belts, 167–68
Van Altena, William, 59 *ill.*
Vanderbilt University, 106
Vatican Observatory, 294
Vaucouleurs, Gerard de, 278
Vela (constellation), 159; pulsar, 160
Very Large Array (radio telescope), 99
Vienna, Austria, 249
Viking missions, 72
Virgo (constellation), 113, 270, 297
Volkoff, George M., 85
Vulpecula (constellation), 140, 142

Wade, Campbell M:, 202, 205
Walker, E. N., 218–19
Waller, Ivar, 52–53 *ill.*
Washington, D.C., 31, 61, 72–73, 117, 164, 176, 225
Washington University, 72–73
Wavelengths, 119, 163
Weaver, Thomas, 236
Weber, Joseph, 222
Webster, Louise, 205
Weight, defined, 44
Weinberg, Steven, 136, 267; *The First Three Minutes*, 136, 267
Weizmann Institute of Science, 31
Westerbork Observatory, 202, 204–05
Westfall, William, 275

ESPECIALLY FOR YOU
FROM WARNER